Stephen Hawking

Stephen Hawking

A MEMOIR OF FRIENDSHIP AND PHYSICS

Leonard Mlodinow

PANTHEON BOOKS, NEW YORK

All rights reserved. Published in the United States by Pantheon Books, a division of Penguin Random House LLC, New York, and distributed in Canada by Penguin Random House Canada Limited, Toronto.

Pantheon Books and colophon are registered trademarks of Penguin Random House LLC.

Library of Congress Cataloging-in-Publication Data
Name: Mlodinow, Leonard, [date] author.
Title: Stephen Hawking : a memoir of friendship and physics / Leonard Mlodinow.
Description: First edition. New York : Pantheon Books, 2020.
Identifiers: LCCN 2019049362 (print). LCCN 2019049363 (ebook).
ISBN 9781524748685 (hardcover). ISBN 9781524748692 (ebook).
ISBN 9780375715365 (open market).
Subjects: LCSH: Hawking, Stephen, 1942–2018. Physicists—
Great Britain—Biography. People with disabilities—Biography.
Classification: LCC QC16.H33 M56 2020 (print) |
LCC QC16.H33 (ebook) | DDC 530.092 [B]—dc23
LC record available at lccn.loc.gov/2019049362
LC ebook record available at lccn.loc.gov/2019049363

www.pantheonbooks.com

Jacket photograph: The Milky Way / Shutterstock
Jacket design by Kelly Blair

Printed in the United States of America
First Edition

2 4 6 8 9 7 5 3 1

In memory of Stephen Hawking

1942–2018

When it's over, I want to say: all my life
I was a bride married to amazement,
I was the bridegroom, taking the world into my arms.

—MARY OLIVER, "WHEN DEATH COMES"

Stephen Hawking

Introduction

I said my last goodbye to Stephen at Great St. Mary's church, a five-hundred-year-old structure in the midst of old Cambridge. It was March 2018. I sat on the aisle, and as he passed we were, for one final moment, in close proximity. I felt as if I were with him again, despite the coffin that veiled him from me and the other mourners and that, after seventy-six years, finally shielded him from the dangers and challenges of the physical world.

Stephen believed that death is the end. We humans create buildings, theories, and progeny, and the river of time will carry them forward. But we ourselves will eventually be left behind. That was also my belief, and yet, as the coffin passed, I felt as if, inside that wooden box, he was still with us. It was an eerie feeling. My intellect told me that Stephen's blip of existence had passed, as would my own in not so many years. Physics had taught me that someday, not just all that we treasure, but all that we are aware of, will be gone. I knew that even our earth, our sun, and our galaxy are on borrowed time, and that when your time runs out, all that's left is dust. Still, I silently sent Stephen my love and my best wishes for the eternal future.

I looked down at Stephen's contented face on the cover of the funeral program. I thought of his strength, of his broad smiles of appreciation and his fierce grimaces of disapproval. I thought of our happy times immersed together in something we were both passionate about. I thought of the rewarding times when we spoke of beautiful ideas, or when I'd learn something new from him—and of the frustrating times when I would try to convince him of something and he wouldn't budge.

Stephen was world famous for stirring up the physics world, for writing about it, and for doing all that from within a broken body. But just as challenging to someone who cannot move, and especially to someone who cannot speak, is to maintain long-term friendships, to develop deep relationships, and to find love. Stephen knew that it was human bonds, love, and not just his physics, that nourished him. And in that, too, he had succeeded beyond reasonable expectation.

Some of the eulogies alluded to the irony that Stephen, who did not believe in God, was having his funeral in a church. To me it made sense, for despite Stephen's passionate intellectual belief that the laws of science govern everything that happens in nature, he was a deeply spiritual man. He believed in the *human* spirit. He believed that all people have an emotional and moral essence that distinguishes us from other animals and defines us as individuals. Believing that our souls are not supernatural, but rather the product of our brains, did not diminish his spirituality. How could it? To Stephen, a man who could neither speak nor move, his spirit was all that he had.

"Stubbornness is my best quality," Stephen liked to say, and I couldn't argue. Stubbornness enabled him to pursue ideas that seemed to be going nowhere, that others rolled their eyes at. It enabled his spirit to dance in the prison of his limp body. Stephen's life had proceeded in violation of all his doctors' predictions, but on March 14, 2018, Stephen's star finally burned out.

Now we were all here to say goodbye. His family, his friends, his carers,* and his colleagues. He was thirteen years my senior, was supposed to have perished decades before, and had been regularly ill throughout his adulthood with potentially lethal lung infections. But in my heart I'd always assumed he'd outlive me.

I came to know Stephen after he contacted me in 2003. He asked if I would consider writing with him. He'd read my books, *Euclid's Window,* about curved space, and *Feynman's Rainbow,* about my relationship with the legendary physicist. He said that he liked my writing, and he liked that I was a fellow physicist who could understand his work. I was stunned. I was flattered. In the ensuing years, he and I would write two books together, and we would also become friends.

The first book we wrote was *A Briefer History of Time.* That wasn't an original work. It was a rewrite of Stephen's famous *A Brief History of Time.* His idea was to make the original more understandable. Kip Thorne, a Caltech theoretical physicist who was one of Stephen's closest friends, once told me that the more physics you know, the less you understand *A Brief History.* Stephen put it a little differently. "Everybody bought it," he said. "Not many read it."

A Briefer History of Time was published in 2005. I was on the faculty of Caltech at the time. Stephen lived in England, but he visited Caltech each year, for two to four weeks. His visits, and our email communication, had been enough for us to complete *A Briefer History.* That, like *The Universe in a Nutshell* and his other books, was based on research he'd done in the 1970s and

* Stephen called his caregivers "carers." Most were not professional nurses.

'80s. But after *A Briefer History* came out, we decided to write *The Grand Design*. This was to be about his latest work, and we would start from scratch, writing about new theories that he'd never before popularized—and we'd be covering some pretty complex issues. Parallel universes, the idea that the universe could have arisen from a state of nothingness, the fact that the laws of nature seemed fine-tuned in just the manner necessary for life to exist. It was clear that this would be a different game. We'd need a lot of face time. And so I started commuting from California to work with Stephen in Cambridge. I kept at it until we finally finished in 2010.

Much of Stephen's career was spent picking up where Einstein had left off. In 1905, Einstein invented what is now called special relativity. At the time he was twenty-five and working on physics as a hobby while keeping up with his day job, analyzing patents. Relativity exposed many of nature's exotic secrets: that the measurement of distances and time intervals is relative, dependent on the observer; that matter is a form of energy; and that nothing may move faster than light. But there was a problem: while special relativity did not address gravity directly, its dictate of a universal speed limit contradicted Newton's theory, in which that force is transmitted instantaneously—that is, with infinite speed.

Einstein struggled with that contradiction. Must relativity be altered? Must Newton's theory of gravity be abandoned? As it turned out, both were necessary. He studied the problem for ten years, quitting the patent office to bounce among academic positions in Bern, Zurich, Prague, and Berlin. Finally, in 1915,

Einstein completed his new theory, general relativity. It was an extensive revamping of special relativity, an extension of that theory in which the effects of gravity are explicitly taken into account.

One of the many ways general relativity differs from Newton's theory is in correcting Newton's tenet that gravity is instantaneously transmitted: according to general relativity, gravity travels in waves in a manner analogous to waves of light—and at the speed of light, thus obeying special relativity's speed limit. Ironically, though the achievement of a satisfactory description of the transmission of gravity was one of the initial spurs to Einstein's development of general relativity, gravity waves were one of the last major aspects of his theory to be experimentally confirmed. For his "decisive contributions" to that experiment, Kip Thorne shared the Nobel Prize in 2017.

Newton had explained why planets orbit and things fall by imagining a force he called gravity. Gravity attracts all matter to other matter, and causes the paths of objects to deviate from their "natural motion," which he declared proceeds along a straight line. Einstein showed us that this is just an approximate picture, that there exists a deeper truth, according to which the phenomenon of gravity is described in a far different way.

According to Einstein, matter and energy do not exert their attraction through the application of force. Instead, they cause space to curve—while the curvature of space, in turn, determines how matter moves and how energy propagates. Matter acts on space-time, and space-time acts on matter. It's that feedback loop that makes the mathematics of general relativity so difficult. To develop it, Einstein had to learn and master a then obscure mathematical field called non-Euclidean geometry— the mathematics of curved space. During the ten hard-fought years it took him to perfect general relativity, Einstein had to engage in repeated rounds of trial and error, postulating forms the theory might take, calculating the consequences of his provisional theory, and critiquing his own ideas.

In ordinary situations Newton's theories provide a good approximation—that's why it was centuries before anyone noticed their shortcomings. But in regions in which speeds are high, or matter and energy are highly concentrated—and hence gravity is strong—one cannot rely on Newtonian theory.

Today special relativity is employed in many areas of physics. But the contexts in which general relativity is required to make sense of things are limited. The two most important have to do with black holes and the origins of the universe. For decades, these subjects both seemed remote and inaccessible to experiment. The early universe was thought to be too far in the past for fruitful study, and Einstein himself dismissed black holes, thinking that they were mere mathematical curiosities, not a phenomenon that actually occurs in nature. As a result, in the half century following Einstein's 1915 paper, those ideas were largely ignored, and general relativity was relegated to the quiet backwaters of science.

What other physicists thought did not deter Stephen. His first text was in fact a tome he coauthored, *The Large Scale Structure of Space-time,* which was largely concerned with curved space and the mathematics that describes it. I'd read a good part of it in college and found it very exciting, a real page-turner, but only if you turned the pages slowly. It could take an hour or more to digest a single one.

Both black holes and the early universe fascinated Stephen, and he made the physics of those systems his main area of research. His early work had a huge influence on others, and led the way in reviving the slumbering theory of general relativity. Later, his discoveries regarding the interplay between relativity and quantum theory helped launch the field now called quantum gravity.

It is to these ideas and phenomena that Stephen devoted his life. He demonstrated their relevance, and he never ceased mining them for new discoveries. By the time he decided to write *The Grand Design,* after forty years of thought and hard work, Stephen believed he'd finally understood the answers to the toughest of the questions he'd had when he started his career— *how did the universe begin, why is there a universe at all,* and *why are the laws of physics what they are*? To explain his answers was our purpose in writing *The Grand Design.*

When you work with someone on a project you are passionate about, you must connect your minds. If you are lucky, you'll also connect your hearts. In working together we became friends. What began as an alliance of intellect grew into a connection of our humanity. I was surprised but shouldn't have been, for Stephen did not just search for the secrets of the universe, he also searched for people to share them with.

As a child, Stephen was bullied by other boys. "He was small and looked like a monkey," said a former high school classmate. As an adult, he was captive in a dysfunctional frame. But he fought the bullying with humor, and his paralysis with inner strength. No one who knew Stephen well was unaffected by his strong personal qualities, or by his scientific vision. In the pages that follow, I'm going to share my experience of working with Stephen and coming to know him as a friend. I hope to shed light on what made him special, both as a physicist and as a person. What was he really like? How did he cope with his illness, and how did his disability affect his thinking? What distinguished his approach to life, and to science? What inspired him, and how did his ideas originate? What were his main scientific accomplishments, and how do they fit into the whole of physics? What is it that theoretical physicists really do, anyway, how do they do it—and why? As I worked with Stephen I gained a new perspective on all these questions, including those about which I had my own initial opinions. My aim, as I recall our time together and recount some of the highlights of his life, is to share what I learned.

I

I'm not a gawker, but when I first arrived in Cambridge, in 2006, I gawked. It was the summer of Stephen's sixty-fourth year, and although many of the details of his life did not match those that would be portrayed in the Hollywood film about him, the details of Cambridge *did* seem a close match to what I'd seen in another film—a Harry Potter movie. Cambridge was Hogwarts. The outer neighborhoods of the city probably have less charm and history, but I rarely ventured far from the "old Cambridge" that Newton knew, a mass of stone streets and buildings that had sprung up in seemingly random places. It's where much of the university is located, mingled with medieval churches and churchyards. It's a place of high walls built centuries ago to protect the students from the townspeople, narrow walkways, and almost-as-narrow brick streets laid down in a jumbled fashion. They were like strands of limp linguine.

The unplanned and irregular layout of the city is understandable when you realize that the university was founded eight hundred years earlier, centuries before René Descartes invented his neat rectangular coordinates. Still, "old" is a relative term: people have actually lived in the Cambridge area since prehistoric times. Today the university is composed of thirty-one semi-autonomous colleges, and more than a hundred thousand people live in the city.

If Cambridge had the Hogwarts look, there was an essential difference. The magic done here was real. There was the courtyard where Newton stamped his foot to time the echoes, and measured the speed of sound; the laboratory built by James Clerk Maxwell, who puzzled out the secrets of electricity and magnetism, and where J. J. Thomson discovered the electron; the bar where Watson and Crick used to drink beer and talk genetics; the building where Ernest Rutherford—the man who unlocked the mystery of atomic structure—conducted his careful experiments.

In Cambridge they are rightly proud of their tradition of science, and they call Oxford, which is more humanities oriented, "that other school." The head of Stephen's department told me that he, like Stephen, had been an undergraduate at Oxford, and his professors required them to write essays on scientific issues rather than just assigning the usual homework problems. He said he tried to assign essays at Cambridge but that none of his students turned one in. These were die-hard science types, and if they were destined to win a Nobel Prize it wouldn't be in literature.

On my visits, Stephen had me housed at the college that he was affiliated with, Gonville & Caius,* in a compound in old Cambridge that dated back to the fourteenth century. The first day of my first visit, I decided to walk from there to Stephen's office. It took only twenty minutes, but the sun beat down on me and I wasn't used to the humidity. Stephen had always appreciated Caltech's Southern California winters. He suffered fewer lung infections out there, and he hated the freezing Cambridge

* Caius is pronounced *keys*.

winters. Now that I was there I realized the Cambridge summers weren't so great, either. The British complain a lot about the weather. They have reason.

By the time I got to the Centre for Mathematical Sciences, the complex of buildings where Stephen had his office, I was ready to be indoors. But it was hard to find Stephen's building. The center comprises seven pavilions arranged in a parabola. Made of brick, metal, and stone, they had large windows and a futuristic, Japanese temple look. I liked the windows, and there were a lot of them. The complex had won some design awards, but the design element I'd have liked most would have been arrows on signs saying "This way to Stephen Hawking."

Stephen's pavilion was adjacent to an older building called the Isaac Newton Institute. Newton's name came up a lot when you knew Stephen. People even compared him to Newton, which is ironic because Stephen didn't like Newton. Newton engaged in many petty squabbles, and was conniving and vindictive when in a position of power. He refused to share credit for any of his discoveries or even to acknowledge that he'd been influenced by the ideas of others. He was also humorless. A relative who'd been his assistant for five years said he'd only seen Newton laugh once, when someone asked him why anyone would want to study Euclid. I'd read several biographies of the man, and though they had various titles, any one of them could have been called *Isaac Newton: What an Ass*.

Perhaps more important than Stephen's estimation of Newton's disposition is that, in high school, Stephen had been bored by the Newtonian physics he was taught. What excites a scientist is discovery—the revelation of a type of behavior that no one has ever seen, or the achievement of an understanding no one has ever had. But since Newton's laws describe the everyday world, and because they are centuries old, there were no surprises in high school physics. In high school, teachers use Newton's laws to describe a swinging pendulum or predict what happens when billiard balls collide. To Stephen, the lesson in that seemed to be *Fun people play billiards; physicists write equations for it*. And so, in those early days of Stephen's education, he had no patience for physics. He liked chemistry better. At least in chemistry, things explode now and then.

Stephen's pavilion at the Centre for Mathematical Sciences housed the Department of Applied Mathematics and Theoretical Physics, or DAMTP, as people affectionately referred to it, pronouncing the acronym as if the P were silent. DAMTP was world famous as Stephen Hawking's university department.

There were only three stories in Stephen's building, and the stairwell wound around an elevator shaft. I went up some stairs to the second floor. The building was wheelchair-accessible. It often angered Stephen when buildings weren't. That was another thing that endeared Caltech to him—when he accepted an appointment to spend a year there in 1974, as part of its welcome the university made the entire campus handicapped-accessible. Such accommodation wasn't required in the United States until the passage of the Americans with Disabilities Act in 1990.

I got to the top of the stairs and turned left, which put me facing Stephen's office door. The door was closed. I couldn't have known what that meant, but I would soon learn. I felt a little nervous about that, and about being there, my first time on his home turf.

As I stepped toward Stephen's door, his palace guard came out to intercept me. Her name was Judith. Stephen's was a corner office, and hers was adjacent. She inserted herself between me and Stephen's door. Judith was formidable. Fiftyish, she was strongly built and had a personality to match. When she was young, she'd spent four years working in Fiji, pioneering art therapy as an alternative to electroshock for the criminally insane. One of the patients assigned to her there had cut off his father's head. Within a few weeks, she had him drawing palm trees with crayons. If she could handle him, she could handle me.

"Are you Leonard?" she asked. She had a powerful voice. I nodded. "Nice to meet you in person," she said. "It'll be just a few minutes. Stephen is on the couch."

Stephen is on the couch. What did that mean? I go to the couch to take naps and watch movies. I didn't think that was what was happening there. But I felt it would be impolite to ask, so I just nodded as if it were a normal thing to be kept waiting while a famous scientist got in his time on the couch.

Though we'd not met before, Judith and I had traded many emails and spoken on the phone. I knew that she was an important force in Stephen's universe. When you requested time with Stephen, it was she who decided whether he was free. When you called, it was she who picked up, and brought him the call (or didn't). When you wrote him, it was she who decided whether to relay the letter, and, if important, to read it to him. The only time I ever heard of someone getting the better of her was when Stephen, while in South Africa, went to see Nelson Mandela, whom he very much admired. Mandela was around ninety then. He wasn't at all tech savvy, and for some reason he was freaked out by the way Stephen's computer spoke for him. He wasn't well, either. He was in frail health. "A little past it" was how Stephen described him, which was ironic because Stephen was having a bad day, too, and almost hadn't made it to the appointment. Judith, though, was part of the entourage on that trip and was keen on meeting Mandela, so she saw to it that Stephen went, and she joined him and his carer for the ride over. But Mandela had his own Judith, a woman named Zelda, and when Stephen and his carer were ushered into a room to meet Mandela, Zelda stepped forward to stop Judith. Too big a crowd for the old man, Zelda had decided, so she wouldn't let Judith through. Zelda had "Judithed" Judith.

My mother used to say, "Where there's a will there's a way." She had a lot of sayings, but this one made sense. Indeed, every security system has its vulnerabilities, and Stephen's was no different. It had a back door. You could circumvent Judith and contact Stephen directly if you knew the email address that he provided to friends and checked himself. The problem was that, more often than not, he would not answer. Even Kip, who'd been Stephen's great friend for decades, told me that Stephen answered his emails only about half the time. Not getting an answer didn't mean that Stephen hadn't read it—but you never knew what it did mean. If he read it, whether you got an answer depended not on how important the issue was to you, but on how important it was to him. At a communication speed of six words a minute, he had to be choosy about doling out answers.

Judith could help you with that, too, if she was on your side. Copy or forward an email to her, and she'd print it out, walk in, and read it to Stephen. And if he was reluctant to answer, she'd push him. Or, if I needed to talk to him, I would call her and she would sit with him and take the call on the speakerphone on his desk. On the other hand, if she'd decided he had better things to do than communicate with you, he'd be strangely unavailable whenever you tried to make contact. After we chatted for a few minutes Judith's phone rang, and she asked me to stay seated in her office while she popped into Stephen's. She emerged in a minute and came for me. His door was now open.

Judith led me in. There was Stephen, sitting in his famous wheelchair, behind his famous desk. He was looking down toward his computer screen. His face appeared young for someone sixty-four. He wore a blue button-down shirt with the top button or two undone, exposing his stoma—the hole at the base of his neck through which he breathed. It looked like a dark red circle of blood the size of a dime. He was very thin, and his shirt and gray dress slacks were correspondingly baggy. The only muscles Stephen could move with regularity were those in his face. His other muscles had deteriorated, so there was a limpness to his body that affected his posture. His head sat unnaturally low between his shoulders as if it were sinking in, and it had a slight tilt. On television this was all part of his look, but viewed in person, it was disconcerting, and though I'd worked with him at Caltech, I wasn't yet used to it. Still, he was an icon, and I felt a bit star-struck—who was I to merit all the time we were going to spend together, to merit him clearing his entire schedule for a week or more at a time to accommodate my visits?

"Hi, Stephen," I said, though he hadn't looked up. "Good to see you. And great to be here. I love Cambridge!"

He still didn't look up. I waited a minute. It got awkward. Then, to fill the silence, I said, "I'm excited to start the book."

As soon as I'd uttered the words, I regretted them. A dumb cliché, I thought, and in any case they didn't fill a whole lot of the silence. Also, what I'd said wasn't technically true. We'd put in some work already, on Stephen's last couple of visits to Caltech. But all we'd done then was discuss what the book would entail. We hadn't yet actually written anything.

I tried to think of something else to say. Something more intelligent. Nothing came to me. Finally, I noticed Stephen twitching his cheek. That was how he typed. His glasses had a sensor that detected the twitches and translated them to mouse clicks, which allowed him to select letters, words, or phrases from lists as the cursor moved around on his screen. It was sort of like playing a computer game. Since he was typing, I figured he was going to reply to my awkward babble. He was going to say something and let me off the hook. After a moment, his computerized voice finally spoke. But all he said was "Banana."

This threw me off. I'd flown six thousand miles, and arrived a couple of days early just so I'd be fresh when I met him, and the only reaction I got was "banana"? What does it mean when you greet someone and he responds with the name of a fruit? I pondered this. But then Sandi, his carer, leapt up from the couch where she'd been sitting, reading a romance novel.

"Banana and kiwi?" she asked.

Stephen raised his eyebrows, meaning yes.

"And tea?"

He again signaled the affirmative.

As Sandi walked around to the mini kitchen behind him, he finally gazed up at me. We locked eyes. Strangely, now, he didn't need to use his words. His expression was warm and happy, and it disarmed me. Now I felt guilty for being impatient with him. He started typing. After a minute or so, the words I had been waiting for finally came out. "Welcome to DAMTP," his voice said.

I could tell there wouldn't be a lot of small talk, and that was fine with me. I really *was* excited to get to work. But just then a middle-aged man walked in. He was a Cambridge professor, a semi-well-known cosmologist. I recognized him, but I couldn't think of his name. Nor was it offered, and of course Stephen didn't expend the energy to introduce us. "I want to talk to you about Daniel," he said to Stephen, ignoring me. "Do you have a minute?"

In the coming years, I would always find this annoying. People would pop in at random times and interrupt us in the middle of our work. "Just a quick minute," they'd always say. But I soon learned that "quick" was a euphemism for "not quick." Once in, Stephen's colleagues usually spoke to him at length.

Though the interruptions bothered me, Stephen didn't seem to mind them at all.

Stephen raised his eyebrows, meaning yes, which meant I'd have to wait. The conversation was mildly interesting for a while. It seems a student named Daniel's stipend had run out, and he hadn't yet finished his Ph.D. But he'd been working diligently and made a good start. Could the department spring for more cash to support him till he got done? Stephen, as head of the general relativity group, was in charge of allocating certain grant monies to students and young postdocs for support, travel, and other needs.

After a few minutes my mind wandered. I looked around the room. The office was more or less a rectangle, with one of the longer sides the one with the door. The side opposite was lined with windows that provided a lot of light and a nice view of the futuristic complex.

Stephen's desk was just to your left as you walked in, situated perpendicular to the windows. The couch was to the right, with its back pushed against them. Behind Stephen was the mini kitchen—a counter with a sink in it and an electric kettle—and a wall of bookshelves above. To the right and left of the door were blackboards covered with equations scrawled by his many students and collaborators. There was also a photoshopped picture of Stephen with Marilyn Monroe, about whom, in his younger days, he'd had a strange obsession.

The office was large for a university office, smaller only than that of the department head. I'd been in executive offices in the business world and in Hollywood, and you could tell, even before you entered, that these people were movers and shakers. But physics is not a money sport, and Stephen's office was modest. If Stephen had been an executive of comparable fame in the corporate world, this office could've fit in his private bathroom.

They were finally winding down. Bottom line, the professor said, would Stephen approve £6,000 for the guy? Stephen typed his decision: "3,000." The professor thanked him and left. Such issues, it turned out, were not uncommon, and Stephen always said yes to the requests because he had such empathy for his students. But he always halved the amount, not wanting to seem like a soft touch. It didn't work. "He's an absolute push-over," Judith would tell me. "And they all know he halves the amount so they ask for double. It's an odd game, really. Played by odd people. No disrespect intended."

By the time the professor was done making his request, Sandi had long since peeled and mashed a banana and kiwi together, and made a pot of tea. I went to sit on the couch for the next ten minutes as she fed him with a tablespoon. The utensil was on the large side, the perfect size for spooning food into Stephen's mouth. One of his carers had come across it in a local restaurant one day, jammed it into her purse, and stolen it. Now they used it at every meal.

The couch, the famous couch, was bright-orange leather and quite comfortable. I later found out that it was where Ste-phen was carried—by the carer on duty and Sam Blackburn, his computer/electronics assistant—when he needed, with the carer's help, to relieve himself. That explained the meaning of *on the couch*. It also made me feel a bit odd when I sat there.

For Stephen, being on the couch took some time. After-ward he might seem tired out, and he'd often follow up with tea, a mashed banana, or both—as he'd just done. When he was on the couch, I would learn, was pretty much the only time Stephen's door would be closed.

I wondered what it was like for Stephen to always be in the presence of a carer in such an intimate situation. I wondered what it was like to *need* others in that situation. To open yourself up to their help, as he had to. I looked over and he was almost finished. Bits of banana and a stream of tea dribbled from his mouth and down his chin. Sandi wiped them with a napkin. To accept that kind of assistance was a bridge he'd crossed many years earlier, and there was no hint that he felt sorry for himself. Instead, he seemed to feel fortunate that he had the people he needed to help him.

We physicists study how systems change in time, but in our lives we cannot presume to have a vision of what is to come. Another thing my mother used to say was, "You never know what tomorrow will bring." She was a Holocaust survivor, and for her this meant that inalterable disaster could always be just around the corner. The message Stephen gleaned from his own history was the opposite. It said that however rotten the hand you'd been dealt, you could make something of it. His illness cut into him at an early age, but though that wound slowly grew, his life did not diminish. On the contrary, it was steadily enriched. On days when I came to work feeling discouraged about something, seeing Stephen always inspired me and kept those relatively minor problems in perspective.

During Stephen's Caltech visits, we'd formulated a detailed "plan" outlining what each chapter would include. We'd created a grand design for *The Grand Design*. *A Brief History of Time* had outlined what we knew about the origin and evolution of the universe in the early 1980s and addressed the question *How did the universe begin? The Grand Design* would be a natural follow-up, updating that answer but also addressing the issues of why there is a universe at all—did it need a creator?—and why the laws of nature are what they are.

In our plan for the book Stephen and I structured a narrative that illuminated those issues. We parsed Stephen's recent work, and all the background needed in order to appreciate its significance, into a set of subtopics. Then we decided how to split up the writing. Chapter by chapter, we'd agreed on the sections we'd each attack. Our strategy was to compose drafts of our assigned topics and trade them via email, then meet, either in Cambridge or at Caltech, to go over each other's work. Then we'd each make revisions, and repeat the cycle.

In some of the passages Stephen sent, I wouldn't be able to figure out what he was trying to say and would have to go back and read his original physics papers to figure it out. Unlike the agreeableness Stephen had exhibited when we worked on *Briefer History,* with this project he would prove ready to debate every point, no matter how small. It was a slow process, like when ants carry bits of leaf across a road to build a fungus farm. In the end, there'd be so much back-and-forth that it would be difficult to attribute a given passage to either of us.

This was the first of those critiquing visits. We worked for several hours, discussing what we'd each written. Talking to Stephen here in England made the American accent in his computer voice seem odd. He'd been born in England, but his voice was from Kansas.

The heat outside intruded into the office. I had grown weary of wiping the sweat from my brow, but for Stephen it must have been worse. I watched a bead form just below his damp, matted hair. It broke loose and rolled slowly down his face, stopping now and then, like a tease. I imagined the little tickle the droplet produced along its trajectory. For me, a quick dab of tissue both obliterates the drop and scratches the itch. But when you can't move, you are doomed to sit there and take it, the barely perceptible but relentless tingle as the bead follows its Newtonian path, an elementary particle of Chinese water torture. Sandi didn't seem to notice. She glanced up at him now and then, but just went on reading.

I wanted to ask Stephen why he didn't have air conditioning, but it wasn't worth the time it'd take for him to answer so I asked Sandi. She answered me at a fast clip, only half of which I could decipher through her thick cockney accent. The gist was, the building had some sort of environmental control system but it wasn't very good. It did things you didn't want, such as closing the motorized window blinds at 5:00 p.m. each day whether you wanted them shut or not, but it didn't do things you did want, like cooling the air. A few years later Sam would secretly rig a bypass that allowed them to control the blinds themselves. Sam was always coming up with workarounds. More important, for me, he always had the inside scoop on Stephen's schedule. But with regard to summer heat, Sam had no solution.

Stephen had requested that a stand-alone air conditioner be installed or that he be allowed to install his own, but the administration wouldn't have it. No one else, they said, had an air conditioner, so why should they make an exception? Yes, why? Perhaps because Stephen brought the university more fame and attention than the rest of the physics faculty combined? Perhaps because it was only through *his* fundraising efforts that the university was able to build the Centre for Mathematical Sciences in the first place? Or maybe because he was PARALYZED. But the bureaucrats didn't see it that way. His fellow faculty members might have adored him, but the clique who ran the place had never been kind to him. To university faculty members, administrators often seem to care only about legal issues, budgets, and fundraising; to administrators, the faculty seems to care only about their research, and in some cases their students. That usually causes tension between the two groups. I'd expected there would be an exception in Stephen's case, but there wasn't.

Stephen could have gone the route they'd eventually take with the blinds and handled the problem himself. But unlike the blinds switch, an air conditioner would be impossible to hide.

On the other hand, the way things worked at Cambridge, people were often told they couldn't have something, or do something, yet if they got it or did it anyway, the administration wouldn't object. Still, Stephen didn't push the air conditioner issue. On some level I think he agreed with the administration—if the others didn't have one, he shouldn't either.

Sandi said she had to go to the loo. Stephen's carers were told never to leave him alone, and Sandi would normally have informed Judith, who would watch over Stephen in the meantime. But since I was around, Sandi assigned me the job. "Just get Judith if there's a problem," she said. "I'll only be a minute."

When I got back to talking to Stephen, I couldn't help focusing on the sweat. I found myself watching the drops collect on his chin until, under their own weight, they'd break off and fall. Hell with it, I thought. "Shall I dab your brow?" I asked. Stephen raised his eyebrows to indicate yes. One of the few muscle movements he was capable of, he used the brow-raise for many purposes—to reply yes to a query, to indicate that he wanted what you were offering, to say thank you. To say no or express displeasure, on the other hand, he gave you a horrible grimace.

I took a Kleenex, reached over, and gently dabbed his face with it. He raised his brow in gratitude. Since he'd liked it, I decided to gave his face one more dab. As my hand approached, his eyes seemed to flash me a *beware* signal. Life gives me a lot of those, and I usually miss them or perceive them too late. That applied here. My hand, it turned out, was moving a bit fast, and my dab was too enthusiastic. His head, limp like that on a rag doll, tilted and rolled toward his shoulder, then landed on his chest in a painful-looking position.

He grimaced. I felt horrified. What should I do? Is it okay to touch him? What else could I do? I reached over and, as gently as I could, raised his head. His brow and hair were wet from the heat. I let go of him. His head started to slide again. I stopped it. I stood there, holding his head up, trying to balance it. His glasses slid toward his cheek. *Beep beep beep beep.* An alarm started up. I'd been nabbed damaging Stephen Hawking.

Just then Sandi returned, and behind her, Judith, responding to the alarm. Sandi set Stephen's head right and adjusted his glasses. With the glasses reset, the alarm ceased. The glasses had a sensor that detected the distance from his cheek, and sent a signal to his wheelchair computer. Its main purpose was to provide him a way to effect, by flexing a cheek muscle, those mouse clicks that gave him the ability to type and select simple commands from his computer screen. It also provided an alarm that went off if his glasses slid too far down. Judith saw that everything was under control and went back to her office. Sandi wiped Stephen's brow. "I'm sorry," she said. He grimaced. She went back to sitting on the couch.

I felt sorry for Stephen because when his brow itched or sweated, he couldn't scratch or wipe it. This was the period in which I often felt sorry for Stephen. I felt sorry that he had a disability that prevented him from doing most of the ordinary things a person does. That he couldn't feed himself or speak or turn the pages of a book he wanted to read. That he couldn't even attend to his bodily needs. That he had so many thoughts and ideas locked in his brain and a huge bottleneck getting any of it out. Over time all that pity would evaporate like one of Stephen's black holes.

2

Before Stephen had his cadre of carers, some of his graduate assistants would live in the house with him and his then wife, Jane, helping to care for him. It was the 1970s. He was wheelchair bound but had some muscular control, and he could still speak, though in a garbled manner. His students would help dress and feed him in the morning and walk beside him as he drove his wheelchair to work. He'd sometimes challenge them by posing little physics problems to solve along the way, a race to see which of them could get the answer first, before they got there.

Stephen and I never raced to solve anything. But when I got to know him better, I, too, learned how to provide a modicum of routine help if he needed it. I learned to gently touch him to wipe his brow or mouth without causing serious head injury or setting off alarms. Sometimes when I tended to him, my mind would jump to thoughts of what his life was like when his muscles still worked and he could coordinate them. We talked about it once or twice.

His attitude surprised me. He wasn't embittered by his disabilities, as some would have been. Here was a man who redefined hardship for me, who every day, every hour, every minute, had to struggle in a way I never did, who had to constantly endure what would for me have been, in turn, embarrassing, humiliating, painful, exhausting, or daunting. No one would have blamed him if he'd felt sorry for himself. We all do at times, and with much less justification. For me, one migraine and I think Job had a good deal. But Stephen faced each challenge and each new day with humor and a positive outlook. It was the perspective of a man who'd found his place in the world and was happy with it.

To Stephen's Cambridge friends he'd always been like that, but they hadn't known him at Oxford. In 1959, aged seventeen and in seemingly good health, Stephen had gone off to a three-year course of undergraduate study there, focusing on natural science, with an emphasis on physics. He felt lonely in his first two years there. He thrived on friendships, and he wasn't forming any. Then, in his last year, he joined the boat club. He became the coxwain, or cox, on the rowing team, and found both friendship and adventure on the Thames. The Thames flows through Oxford, and Cambridge has the Cam; and rowing had a long tradition at both universities. It was also the practice at both schools for all the popular kids to be on the team. For Stephen, the rowing team was a social club.

As cox, Stephen had the job of controlling the boat's steering and speed. Sitting at the stern of the boat, he'd do some of that directly, but also bark orders to the others. He'd always been awkward and puny, teased, and poor at sports. But now he was in charge. He was perfect for the role—light, so he didn't add a lot of dead weight, and loud.

Despite the rowing team, Stephen was bored with life at Oxford. He was required to attend a number of lectures each week and a tutorial in which the week's homework problems were discussed. He found those problems "ridiculously easy."

He spent little time on them, or on any other schoolwork, preferring to listen to classical music or read science fiction novels. He had no ambition, no goals, no direction. And like most of his fellow students, he drank too much. That was before he went to Cambridge for graduate school, before he got his death sentence, and before he found physics.

Another Cambridge visit. We had just started our work for the day and I'd put a few pages in front of Stephen, on the stand that sat on his desk. I'd written them that morning, and Judith had printed them for me. We began reading them to ourselves. Stephen had lost a little control over his eyes, and so his reading had slowed a bit. Also, when Stephen read a page, knowing that he could not easily revisit it, he tended to study it very closely. When he'd finish, he'd raise his eyebrows, and I'd turn to the next page. When we got to the end, I'd turn back to the first page, and we'd start over, each of us making comments on this round.

For a couple of years at Caltech I taught science writing instead of the usual math class, and relied on physics for some useful analogies to the writing process. In physics we have different theories for different size scales. We generally use quantum theory for the atomic and subatomic scale; Newtonian physics as an "effective" or approximate theory that works well for the everyday scale; and general relativity for the cosmic scale, which is dominated by the force of gravity. You can analyze writing in an analogous way, I taught. There are the word choices, the sentences, the paragraphs, the chapters, the book. You have concerns and tools at each level, some for the big picture and others when you analyze the nitty-gritty.

Given Stephen's limitations and the effort required for him to communicate, you'd think that, in our work, he'd have confined himself to just the most important points and large-scale concerns. But that wasn't the case. He seemed to find no point too small or minor to debate, and no discussion too long to be worthwhile. Sometimes we'd discuss every sentence on a page. The end of his life may have often seemed near, but he didn't let that rush him.

In all the years I worked with Stephen, if we were having a disagreement, he never grew weary of the work involved in typing his responses. Even though I can shoot words out of my mouth at a rapid clip while his sentences had to be painstakingly crafted, it was he who sometimes wore me out, and never vice versa. He worked as if we had forever. Sometimes I'd try to push him by reminding him that we had a deadline. He didn't care. And when we missed the deadline, our publisher would accommodate us with a new deadline. Stephen once told me it'd be fine if the book took us ten years. I told him that if we ever reached our tenth anniversary I'd give him a cake, wish him luck, and then fly home and let him finish the book himself.

This was early in our collaboration on *The Grand Design,* and I hadn't yet learned to give myself up to his perfectionism. He was a man on a mission. He hadn't always been obsessed with doing things just right, however. On the rowing team, for example, though he was perfect for the position of cox, he wasn't good at it. And he hadn't minded. That was part of the problem. To him, rowing meant adventure and companionship; he wasn't in it to win trophies. His rowing coach remembered that he lacked ambition and often seemed distracted. He criticized Stephen for being reckless in races, often trying to shoot through narrow, impassable gaps. But to Stephen, the chance for recklessness was just the point. Why do this otherwise?

Stephen's coach had another issue with him. Under Stephen, not only did the team not do well, it finished its races with damaged oars and pieces of the boat knocked off. Once, Stephen even piloted the boat into a head-on collision. He seemed proud of it. Back then, he had a young body, and unlike the boats, it could take a beating. He was not even twenty and was enjoying his physical power, not thinking what a gift it was. Like most young people—and like a lot of older ones, too—he thought his health, his strength, his intellect, his energy, were eternal.

In physics, a theory is either true or it isn't. A philosopher would say we physicists use that word sloppily, because all "truth" in physics is provisional—you can never be sure that a future experiment won't come along and contradict your hitherto perfect theory. But what I mean by "a theory is either true or it isn't" is that, in fundamental physics, at least, there is no "almost right." If a theory fails even a single obscure test, by even the tiniest amount, we say that the theory had been *falsified*. It cannot represent the true laws of nature.

A falsified theory might still be useful. It might hold true in some limited realm—where distances are large or maybe where they are small, or where speeds are low or gravity is weak. Such effective or approximate theories have a place in practical applications such as solid-state physics, quantum computing, or stellar physics. But if a theory provides a false prediction, no matter how numerically insignificant, those who seek the fundamental laws know they must keep looking.*

The goal of those working in fundamental physics is to

---

* Assuming that the discrepancy is not due to experimental error or a shortcoming of the approximation method employed to derive the prediction.

find theories that hold without exception. But if a supposedly fundamental theory is found to have a flaw, theorists don't mourn—we find that exciting. We set out looking for the next theory, one that explains all that was explained by the prior one while also passing the test in which the old one failed. That next theory might be a modified form of the old theory—as when, in 1998, the so-called standard model of elementary particle theory was altered when it was discovered that neutrinos have mass. Or it could be a completely new theory, as when Newton's laws of motion and gravity were replaced by those of quantum physics and general relativity.

The parade of ever-improving theories ends, in principle anyway, with the so-called theory of everything. However, the verdict on whether a theory of everything actually exists, and if so what it might be, isn't yet in. To create such a theory was Einstein's central goal in the latter part of his life. He called it a unified field theory. You'd think that if anyone could have pulled such a theory out of his hat, it would have been Einstein. But what he mainly succeeded in doing in those last decades of work was to alienate himself from the mainstream of physics. "The current generation sees in me," he wrote, "both a heretic and a reactionary who has, so to speak, outlived himself."

That was fine with Einstein. He felt that he'd achieved plenty of fame and had earned the right, later in life, to engage in quixotic pursuits. And so he ignored the advice of virtually everyone, and kept at it. Like Stephen, he was stubborn. And sure enough, a few decades after Einstein's death in 1955, the quest for a theory of everything finally became fashionable.

Most discussions of a theory of everything overlook the fact that physicists had already believed they had a consistent description of all physics phenomena in the latter half of the nineteenth century. That's when James Clerk Maxwell created a theory of the forces of electromagnetism. Along with Newton's

Law of Gravity, the two theories provided an account of all the (then) known forces in nature. When physicists added to the mix Newton's Laws of Motion—which describe how objects move when acted upon by those forces—the collection of theories seemed to be sufficient to account for all processes in the universe. At least in principle: to apply the theory you still had to solve the equations that described the process of interest.

That's a big admonition. Without a solution to its equations, a theory is just a framework of principles and methods. Given a physical system (the electrons and nucleus of an atom, the solar system, and so on), a theory will provide equations whose solutions describe how the properties of that system evolve in time (the radiation of an atom, the orbits of the planets, and so on). But we usually can't solve those equations, so most results in theoretical physics depend upon the use of appropriate approximations. That's part of what makes physics an art as well as a science.

Physicists in the late nineteenth century were so confident about their theories that in April 1900, Lord Kelvin, one of the most famous scientists of the day, gave a speech about the future of physics in which he implied that all that was left to finish the "physicist's job" was to clear up two clouds in an otherwise clear blue sky. One of those clouds was an experiment on the speed of light by American physicists Albert Michelson and Edward Morley. The other was a phenomenon called blackbody radiation. These were just little anomalies that we'd soon explain within our existing framework of ideas, Kelvin believed. As it turned out, instead of viewing those issues as two little clouds in a blue sky, it would have been more accurate to envision two massive icebergs confronting a ship in the ocean.

To explain the Michelson-Morley experiment, the theory of special relativity needed to be invented, which wouldn't happen until Einstein did it in 1905. To explain blackbody radiation, quantum theory needed to be created, which would require the work of a number of brilliant physicists over several decades, from 1900 to 1925. Together, those new theories sunk the ship of the Newtonian Laws of Motion, which were the foundation for all physics in Kelvin's day, and had been for centuries. Never again would those laws be thought of as representing fundamental truth.

Kelvin wasn't just wrong in dismissing those two anomalies; he also missed spotting a third one. In the mid-nineteenth century, it had been discovered that Mercury deviated a small amount from the predictions of Newton's Law of Gravity. The error was tiny but it was real, and it reflected a flaw in that law. As it turned out, the discovery of that minute deviation foreshadowed a third physics revolution—Einstein's 1915 theory of general relativity.

With the advent of special and general relativity, and quantum theory, Newton's laws of motion and gravity had all been replaced. Maxwell's theory of electromagnetism fared better in that period of upheaval. His laws required revision to fit special relativity and the quantum context but remain recognizable to this day. The lesson is that, in physics, when a theory is sick, that is sometimes signaled by what seem to be mere anomalies, deviations that you mistakenly believe will turn out to have an innocuous explanation.*

The same proved to be true of Stephen's health. In that last year at Oxford, he had found himself having problems with his coordination. He was growing a bit clumsy and slurring his speech just a little. Those two seemingly minor maladies were

---

* Dark energy and dark matter represent two unexplained anomalies in today's physics. Will these eventually be explained within our current framework of theories, or by amending them, or will they require a completely new approach? No one knows.

tips of an iceberg of a fundamental sickness within him. He looked like a healthy teen, but the truth of his body was not health. It was illness.

Unlike Lord Kelvin, Stephen did not ignore the icebergs. Suspecting that something was seriously wrong, he went to the campus infirmary. A doctor there examined him, and then sent him on his way with a prescription to "lay off the beer."

It was late afternoon. Carol, Stephen's carer, was spooning him some mashed banana and coffee. He looked pale, and kept closing and opening his eyes. We usually worked till at least seven, but the eye action meant he was tired. Carol finished feeding him and walked around to turn on the electric pot to make herself some coffee. It was only instant coffee, but the aroma always filled the room.

We didn't always work on the book in the order in which the chapters would appear. We'd bounce around sometimes, if what we wrote later demanded some setup earlier in the book or if one of us simply had a new idea for something we'd already completed. In this case Stephen had suddenly decided that we should go back and start the opening chapter, "The Rule of Law," differently. He wanted to begin with a myth explaining some natural phenomenon that would illustrate why people turn to mythology for their explanations of what they don't understand—and how they might eventually have noticed that their explanation was wrong. But after searching for an hour online, we hadn't found anything appropriate.

We decided to leave the myth as a TBD and to continue refining our ideas for the rest of the chapter. Since TBD meant "for Leonard to figure out," I had plenty to do before coming in the next day. I suggested that he might want to quit early while I went to some pub and worked on it. Before he could answer, Margaret, another carer, walked in. She wasn't on duty but had just dropped by to say hello, as she sometimes did. My suggestion to Stephen went unanswered as she took over the room. She wasn't one who worried about what she might be interrupting.

Twentysomething, reddish-blond, slim, and attractive, Margaret had reputedly once given Stephen a nude portrait she had had painted of herself. She would eventually resign after accidentally pushing Stephen's wheelchair into a doorway and breaking his leg. He wouldn't hold it against her—he seemed to take it with the same generous patience with which he (sometimes) endured the misguided idea of a graduate student. Still, Margaret left, and years later, when I mentioned Margaret's name, Stephen said, "I miss her."

On this beautiful day, Margaret was in classic form, determined that I would have some local fun. When I told her I'd been thinking of quitting early, she announced that this was the time to enjoy some tourist activities. Why didn't she take me "punting on the Cam"? A punt is a rectangular flat-bottomed boat about twenty feet long and three feet wide, with a small platform at the back and a hull that rises just a bit above the waterline. The "punter" stands on the platform and propels the boat by pushing against the riverbed with a long pole, while passengers sit on blankets and lean against one of the back supports that run across the bow. Margaret proposed to serve as punter.

Stephen seemed to suddenly perk up. He raised his brow. He wanted in.

That surprised me. I didn't know it then, because this was before I knew of Stephen's rowing days at Oxford, but Stephen loved being on the river. However, I had read up enough on punting that I knew such a trip could be dicey for Stephen. If the punter lost balance and fell off while trying to push the wobbly craft along, that could tip the boat over. And I'd read that occasionally passengers fell in when boats collided or when they lost their balance while climbing on and off.

For us such mishaps could be awkward and embarrassing, presenting no hazard greater than getting wet. For Stephen, they could be deadly. His lack of muscle control—and hence of muscle usage—had many ancillary effects, one of which was that his bones were weak and brittle because they had not been strengthened by the pulls and tugs they were designed to experience each day. That's why his leg broke when Margaret bumped it, and why it was inadvisable to carry him long distances, as we'd have to do to get down to the boat landing.

Just as important, when Stephen was not accompanied by his computer he couldn't type, so he couldn't communicate his often acute needs. For example, he occasionally had difficulty breathing because the stoma in his throat needed to be suctioned out, and without his computer's voice generator he had no way of asking for that. There was also the possibility that one of us might slip while getting onto the punt. Worst-case scenario, he might fall into the water, in which case there'd be no saving him. He had to know all this, but it didn't deter him. Once I got to know him better, I realized it probably attracted him. Danger seemed to make him feel alive. In life as in his physics, he liked to take chances.

A half hour later, Stephen's van pulled up at the top of a long flight of stone steps leading down to the river. As the car's lift lowered Stephen in his motorized wheelchair down to the street, Carol snatched the large black bag and the smaller silver one that held Stephen's medical gear. Margaret had somehow procured a bottle of French champagne and some strawberries, the classic punting picnic.

Carol and Margaret lifted Stephen out of the chair.

"I can carry him," I offered.

I was, after all, twice their size, and it was quite a long walk down the uneven steps. Later in our relationship, I would occasionally help carry him, but this time Carol chuckled and said they wouldn't endanger Stephen by putting him in *my* hands. Then she and Margaret each grasped one end of Stephen's limp body and proceeded down the stairs, trailed by me carrying the gear and Carol's pink purse.

Neither woman supported Stephen's head, which wobbled freely as they walked. That's when I learned that caring for Stephen was not a science. I recalled when I'd pushed his head over: the beeping alarm, my horror and Stephen's annoyance. Now his head was swinging like a pendulum and everyone was smiling. All I could think about, watching that, was neck pain. I looked for a grimace on Stephen's face, but I couldn't detect one. Of course it was difficult to tell, since I was a ways back and his head was in motion. I thought about saying something, but these were his trusted carers. They'd worked with him for years. I decided to keep quiet. I'd let them do their job and I would do mine—carrying the pink purse.

The Cam is the main river flowing through Cambridge. Lined with lush greenery and old college buildings, it's shallow but navigable by small boats and rowing craft. At various points, eight of the thirty-one Cambridge colleges back onto the river, providing punters with picturesque views of their grand buildings and grounds. It was great fun but not the most comfortable experience. The hard seats of the boat were just a few inches above its deck.

The carers stepped into the boat first, carrying Stephen. Carol sat with her back resting against the bow, legs crossed, opposite the punting platform, which was aft. They positioned Stephen resting on Carol in a half sitting, half prone position, facing the stern, with Carol cradling him.

Though Stephen couldn't speak, he wasn't exactly passive

in this process. He was rarely passive. He indicated whether he wanted to be moved over a bit this way or that by moving his eyes left or right. With grimaces he communicated that the position wasn't quite right, and with raised brows and smiles indicated when it was getting better. When he was finally positioned to his satisfaction, I stepped on board, but the boat was bobbing and I lost balance. For a terrible moment I thought I'd fall on Stephen, but I bent my knees and righted myself. As Stephen watched me stumble, he had a big smile on his face. He'd made it into position easier than I had. I felt sheepish, and it wouldn't be the last time this happened. I'd feel bad for Stephen because of this or that, and then he'd prove to be able to handle it just fine, maybe even better than me.

And so in a bit we were floating along. Carol would turn Stephen's head first left, then right, so that he could gaze at the sights while Margaret propelled us and I fed Stephen bits of strawberries and sips of champagne.

In Stephen's day, undergraduate study at Oxford was not very demanding. The hour or so each day that he studied, he told me, was typical. When he said this, with a smile, I was appalled. Here were not just the brightest kids, but privileged ones, and they were wasting their time and the school's resources. Their attitude was, if you had to work a lot, you must not be very smart. I'd gone to good schools, too. We partied plenty, but we also worked plenty—hard and for long hours.

In Stephen's case, the goofing off was his natural response to not having a passion for anything. He even considered going into the civil service upon graduating. He went through a series of interviews, reporting his job preference as Ministry of

Works, which at the time looked after public buildings. He also expressed interest in the position of House of Commons clerk, even though he had no idea what a House of Commons clerk did. In the end, Stephen's apathy saved him from any serious chance at ending up in one of those jobs—when the day of the civil service exam came, he forgot to go.

In the fall of 1962, after graduating from Oxford at age twenty, Stephen moved to Cambridge to pursue a Ph.D. in physics. His first term there did not go well. His undergrad days had been fun, but that was the past. In graduate school you can't lie on the sofa for months on end, relegating all study to the hour before dinner. And the shoddiness of his preparation at Oxford now became obvious when he found himself lost in the lectures he attended at Cambridge. When Stephen went home for Christmas that year, he was close to flunking out.

Stephen's history of wandering through life must have alarmed his mother. So did his physical condition—he was growing increasingly clumsy. She took him to the family doctor. The family doctor sent him to a specialist. The specialist sent him to the hospital for testing. His family had arranged a private room for him there, but he refused it, due to his "socialist principles." Over the course of a two-week stay, doctors took a sample of muscle from his arm, stuck him with electrodes, injected him with a radioactive fluid, and made a variety of measurements. But like the Oxford doctor, they sent him away without a definitive conclusion. All they said was, it's "not multiple sclerosis." They suggested that Stephen return to Cambridge and continue his graduate work.

In Cambridge, Stephen's symptoms continued to get worse. He felt that he was dying, and found it hard to focus. Finally, the hospital doctors put together his test results and gave him a diagnosis—amyotrophic lateral sclerosis (ALS), a progressive degenerative disease of the motor neurons.

"I felt somewhat of a tragic character," he said, describing his reaction to the diagnosis. That was in early 1963. Stephen had just turned twenty-one.

Like mold in the walls of a house, once it starts—often in the lower extremities—ALS gradually spreads until it takes over. It kills you by destroying your motor neurons, which reach from the brain to the spinal cord, and from the spinal cord to the muscles throughout the body. When the motor neurons die, the ability of the brain to initiate and control muscle movement is lost. Only voluntary muscle action is affected.

In Stephen, the disease more or less started at the legs and worked its way up. When he lost control of his thighs he could no longer stand. When he lost control of his trunk muscles he could no longer sit unsupported. When it attacked his chest, it became difficult to breathe. In 1985, when he was in his forties and had had the disease for twenty years, he had to have the tracheostomy that created his stoma and took away his ability to speak. His mind worked perfectly, but its carrier had become inert.

Death from ALS almost always comes within two to five years of diagnosis. In one out of twenty cases, a patient will live twenty or more years. Stephen lived with it for fifty. When he was first diagnosed, however, as far as he knew he'd be dead in just a few. He expected that he'd eventually suffocate, and that it would be sooner rather than later.

With death apparently imminent, Stephen went through the stages of grief that are part of the experience of having a terminal illness. He eventually sank into a deep depression. He'd shut himself away in a dark room and play Wagner at full blast. It was a remnant of his childhood days, when his parents would also play it at high volume on their record player at home.

At some point Stephen started having dreams about death. In one, he said, he dreamt that he was going to be executed; in another, a recurrent dream, he dreamt that he would sacrifice his life to save others. He started thinking about the meaning of those dreams. With an irrevocable death sentence, he began to question how he spent his time. What could he do with his remaining years or months that would be meaningful? Could he have a passion for *anything*?

If what seems like a little problem can turn out to be major, another paradox of life is that what seems bad can turn out to be good. That was true in the case of those physics icebergs—they may have sunk the ship of Newtonian science, but they guided the way toward the new physics. Stephen told me that his illness guided him to something new, too.

"We all know we will die. For most, that is an abstract thought. It is not abstract for me," he said. It inspired him to value each of his remaining days.

I could see that during our punting trip. Most of us amble through our lives without much sense of urgency. The norm in our society is to pursue career, money, material possessions. We worry about whether our clothes look right, if the car needs washing, whether it's time to trade our cellphone for the latest model. We fill much of our time with stuff that doesn't really matter. Believing that death was imminent steered Stephen toward a richer life in the days he had left. He turned his focus to things others often take for granted—not just the work he developed a passion for, but the people he was close to and the natural world that surrounded him. When Stephen gazed out at the river, I could see in his eyes how important it was to him. It seemed to affect him deeply. I could also see this when I watched him look up at the stars at night. Knowing that death could come at any time, he became conscious of the beauty of each moment of life.

After his diagnosis, it took about a year of intense emotional struggle for Stephen to come to grips with his fate. In defining an ever-growing universe of physical activities he could not do, his disease magnified the value of the mental activities he could. It left him the choice of wasting away in spirit as well as body or finding a world of the mind in which he could still function. Where some in his situation would have found God, Stephen found physics. He decided to finish his Ph.D. He found, to his surprise, that he liked the work.

Ancient philosophers and modern psychologists stress that happiness comes from within. You can be as happy living in a cave with no possessions as you can with a Ferrari and a fancy job. Maybe happier. The deterioration of his body caused Stephen to seek fulfillment internally, within the actions of his mind. Until then, his mind had been dormant. It might have sputtered into operation occasionally, as when he had to pass an exam, but it quickly returned to sleep mode. Then came Stephen's diagnosis. It awakened him. It became an inspiration. And so, as his body withered, his mind blossomed. He began to think about what is important in life. He began to seek meaning, to ponder existential questions about the universe and our place in it. He also became anxious to start a family. And once he achieved some measure of fame and his influence grew, he began to actively seek ways he could help others who suffer, especially those with disabilities.

All the times Stephen and I talked, he never showed a sign of feeling sorry for himself. I asked his buddies Kip Thorne and astronomer Martin Rees, who'd both met him soon after his diagnosis, and they hadn't seen any evidence of self-pity either. Stephen would be dying slowly as the illness insidiously spread, but he wouldn't be mourning his losses. Being with Stephen made all his friends wonder, *Are we awake to our own potential?*

The punting trip had taken only an hour or two, but the normality of accompanying Stephen on this routine tourist activity opened my eyes to the way he chose to live. We returned to the boat landing with Stephen intact, and I seemed to have been the only one who'd considered that it might not end that way.

Back at the van, Stephen's carers began the long process of lowering the van's wheelchair ramp, getting the chair out of the van, seating Stephen in it, strapping him in, driving the chair back onto the van, and fixing it in place. In the office, Stephen had looked pale. Now his color had returned. I, on the other hand, was feeling weary. I was ready to go back to my room for a nap, after which I could start scouring the Internet for the myth we'd decided to include in our opening chapter. But Stephen asked me to come home with him. There was still an hour or two before dinner, he said. He wanted to keep working.

3

Stephen lived in a house on shady Wordsworth Grove, a quiet street just a short walk outside the old city and close to his office. It was a two-story brick dwelling with a black shingle roof. Looked a bit like a Swiss chalet. It was quite a step up from the home of his childhood. That was a house with broken-down furniture, peeling wallpaper, and no central heat. His parents weren't poor, but they were frugal. This house, though, was top-notch, inside and out. The grounds were small but verdant, surrounded by a wood fence covered with lush ivy and other shrubbery. All you could see, looking from the street, was the upper floor. That the house had an upper floor surprised me, since Stephen couldn't get up staircases. From what I'd heard, the woman he was married to at that time, Elaine, liked it that way, liked having her own parallel but inaccessible space. She might have been up there that evening, but she didn't join us. From what I'd also heard, that was not unusual.

Though Stephen said he'd invited me home so we could keep working, it was a long time before we got around to that. First, he'd had to relieve himself. Then came a discussion about what dinner would be. Joan Godwin, an elderly white-haired nurse who had been with Stephen for decades but no longer did carer duty, was there to do the cooking. Joan knew everything about Stephen, including what he needed and how to please him. She could have been his big sister, and was always ready to share with me her insights about him.

After the discussion with Joan, Stephen had teatime, and then he took his vitamins, a dozen or more pills to swallow. After that, he initiated some conversation, which included inviting me to stay for dinner. By the end of it all, an hour had passed. He didn't mind the delay. I had the impression he'd asked me over just to keep him company. It wasn't the last time I'd have that impression. Sometimes, if we hadn't planned to work on a Saturday, he'd summon me at the last minute. I'd think he had a hot idea to discuss, but when I'd get there we'd mostly chat or watch the news on TV. On weekdays, not long after that first dinner, he started inviting me for dinner almost every night. It became the norm, and we'd have to talk about dinner plans only if we were going to do something different, to eat out rather than at his house, or if he or I couldn't make it that night.

I liked hanging out with Stephen, but that night I was happy when we finally got down to work. We had a couple of short exchanges, and then I had the chance to ask a question I deemed important. I was hoping Joan would take her time with the food so we could go on about it for a while. Or that she'd burn the food and have to start over. My question had to do with the myth Stephen wanted me to find for the opening chapter. If I didn't fully understand his thinking, I might waste hours barking up the wrong tree. Stephen had just started typing his response when Joan walked over, carrying a big plate of steaks and mashed potatoes. He ignored her and the meat and continued typing. She went back for the gravy. When Stephen was done typing, his eyes turned toward me while his computer voice said, "Please pick the wine." That was the end of our work for the evening. The myth would have to wait.

Joan showed me where the wine was kept. Stephen had a closet full of it. Mostly red. I think many of the bottles were gifts. Some looked expensive. French, with classifications like Grand Cru on the label and vintages going back to before my kids were born. Stephen always asked me to select the bottle. He seemed to assume I knew something about wine, maybe because I am from California, but I didn't, so I'd pick one at random or choose a vintage that reminded me of some past event. A 1998 Bordeaux? That's the year the French won the World Cup. Let's try it.

Dinner at Stephen's was always pleasant, but we never talked work during our meals. Due to the way he had to type out words with his cheek muscle, he couldn't speak much while eating, anyway. Sometimes words came out inadvertently, just gibberish. But his carers always tried to keep a conversation going. In this case, that was Bella, a Czech woman who'd come on duty shortly after our punting trip. Bella complained about the mushrooms in the gravy. Bella did not like mushrooms. Stephen attributed that to her having grown up in Eastern Europe. There weren't any mushrooms available in the communist times, he told her. Though she'd been a young child when the Soviet Union fell, he said the same thing every time Bella turned down a mushroom.

While Stephen was more focused than most on his work, he was also focused on developing friendships. This wasn't easy, given all the constraints his sickness imposed. Physical limitations. The time required for his daily medical and personal care. The awkwardness of his situation. He couldn't exactly decide to casually chat up a stranger. Despite all that, he managed to socialize a fair amount. Back in his heyday, before his body began its downhill trajectory, he'd been unstoppable.

If Stephen's illness changed his life, so too did a train ride he took in 1963, the same year he learned he had ALS. He'd stepped onto the railway platform at St. Albans for the short trip down to London, as he'd done so many times before. This time, however, the workings of chance launched him on a far more significant journey—with the woman who would become his wife. Her name was Jane Wilde.

He'd met Jane before the railway encounter, at a New Year's party, and just after that had sent her an invitation to his twenty-first birthday celebration, on January 8. At neither event had they talked much. Not long after that she'd heard he'd been diagnosed with some paralyzing disease and didn't have long to live. More than a month passed. She hadn't heard from him, and that seemed to be the end of it. It probably would have been, but for their random meeting on the platform.

They sat together on the way down. It was an easy ride, taking only about a half hour, but it gave them, finally, a chance to chat. Thin and disheveled, Stephen had long brown hair that hung over his glasses. He was just beginning to study cosmology at Cambridge. Jane was eighteen and just finishing high school in St. Albans. She didn't know what cosmology was. She told Stephen that she was sorry to hear about his illness. He wrinkled his nose and changed the subject. They talked about this and that for the entire ride. As they neared the station, he told her that he came to London quite often on weekends. Would she like to go to the theater with him on one of those visits?

The night of their date, Stephen took her to a fashionable Italian restaurant in Soho and then to a play at the Old Vic. It was a pricey evening. So pricey that he ran out of money, and she had to pay for their trip home to St. Albans. "I'm terribly sorry," he said.

It wasn't long before he asked her out again, to a fancy college ball. This time he picked her up in his father's old Ford Zephyr. It was a big car, and he handled it fast and recklessly, just as he'd do years later with his wheelchair—when he was still well enough to operate it. Maybe he knew no fear. Maybe he figured he had little to lose. Jane knew fear, and had plenty to lose. She was terrified. "I scarcely dared look at the road in front," she'd recall. "Stephen, on the other hand, seemed to be looking at everything except the road."

The ride wasn't a great start to the night, but Jane tried to focus on the fun ahead. The ball was really a number of balls, dances held simultaneously in various rooms and halls at many different colleges in old Cambridge. The partying went on all night. Due to the progress of his sickness, Stephen was a bit unsteady by then and told her he didn't dance. "That's quite all right," she said. "It doesn't matter." But she was lying. It did matter.

As the night went on they flitted from venue to venue. A Jamaican steel band played on a lawn. A string quartet graced a wood-paneled room. A cabaret played on a distant stage. As they walked between the venues, the sounds blended into each other. There were fine buffets everywhere, and the champagne flowed. They eventually stumbled upon a cellar illuminated only by faint bluish lights. A jazz band was playing, and partyers packed the dance floor. Jane wanted to join in, and this time Stephen let her talk him into it. They moved together to the music until the band stopped playing. When morning came, they sank deep into armchairs in a room at Stephen's college, Trinity, where Newton had worked. They slept for a while.

Jane had had a wonderful night, but when it was time to head home, she remembered being driven in, and it brought her back to earth. She didn't want to go through that ordeal again. She suggested that she take the train home. He was a gentleman, however, so that was a hard sell. Should she make a scene, or just shut her eyes and go along? She chose politeness over comfort. By the time they pulled up to the gate at her house she was a wreck, and ran from the car with barely a goodbye.

Once again that might have been the end of it, but Jane's mother had been watching and was appalled. How rude of her not to invite him in. Jane thought better of it and ran back down to the gate. Their house was on a steep hill, and Stephen had taken the parking brake off before starting the engine. As he fumbled with the key, the car was rolling away. She was glad to be on the outside. Stephen saw her and jammed on the brake.

They had tea in the sun by the garden door. He was attentive and charming as they talked and laughed about the previous night's events. As they sat together, the thought of that alarming ride home began to be a bit more tolerable. It seemed part of the big adventure they'd just had, which she could have more of if she saw him again. She decided that she liked him, strange and reckless as he was. She liked him a lot.

A couple of years later, they married. Over the next three decades she'd love him, watch his honors and triumphs accumulate, appreciate his courage, genius, and humor, devote herself to him, make him a home, pay his bills, bear and raise his three children, and, eventually, feed him, dress him, bathe him, and sit with him through his many hospital visits and near-death experiences. In the process, over time, she'd lose her own identity. And with it, her self-worth. *Who am I?* she'd wonder. *Nobody?*

No matter who you will turn out to be, or how much you know, when you first arrive as a physics graduate student, you start at the bottom. The classes you took as an undergraduate, and even those you will take at the graduate level, are important, but they're just background. They leave you in the position of a builder who has studied buildings, but never built anything. To earn a Ph.D. in theoretical physics, you have to build a structure of your own. Or add onto an existing one. Or find one that needs fixing, and fix it. It is only after you've done that once, or ten times, that you are really a theoretical physicist, and understand what it means to be one.

In most Ph.D. programs, sometime in the year or so after you arrive you have to convince someone on the faculty to be your "advisor"—your mentor and supervisor as you engage in your first construction project. At Cambridge, things worked a bit differently. When Stephen applied for admission, he had to apply to work with a specific person. He asked to work with Fred Hoyle, the most famous British astronomer of his day. Although Cambridge accepted Stephen, he was told that Hoyle already had too many students, and he was instead assigned to another theorist, Dennis Sciama.* Stephen had never heard of Dennis Sciama.

Having the right graduate advisor is important, not just because you want to click with your mentor, but also because your road can be rocky if your interests don't align. Your most basic choice is whether you want to be a theorist or an experimentalist. Most physicists, by far, choose to be experimentalists. That's necessary because it takes many more scientists to build

---

* Pronounced *she-ama*.

the apparatus needed to test a theory than it does to come up with one, and so the demand is far greater. Usually, by the time you enter graduate school, you know in which of those camps your heart lies. But that's just the beginning.

Physics is a vast science, comprised of a great array of specialties and subspecialties. Some are concerned with uncovering the fundamental laws of nature. Others focus on applying those laws to specific phenomena or systems. In optics, for example, the basic laws of electromagnetism are applied to study the behavior of light and how it interacts with matter. Nuclear physics focuses on understanding the interactions of the protons and neutrons inside the atom. In quantum information, the basic laws of quantum theory are applied with the goal of creating superpowerful computers.

The study of the fundamental laws, in contrast, rests on just two main pillars. One, general relativity, is a theory of the force of gravity alone, and how matter moves in response to it. But in addition to gravity, there are three other forces in nature—the electromagnetic force and the strong and weak nuclear forces. These forces have no place in general relativity. They, and their effects, are instead described by a theory called the standard model—the second pillar of the fundamental laws.

The standard model is a quantum theory, which is a type of theory, one that incorporates the quantum hypothesis discovered by Max Planck in the year 1900. That hypothesis states that certain quantities such as energy can only take on discrete values. If in Newtonian theory energy was continuous like water, in Planck's theory it came in distinct subunits, like the tiny particles that make up flour. In quantum theories all properties of particles, fields, and universes become "fuzzy" and probabilistic. Theories that don't incorporate these features are called classical theories, even if, like general relativity, they are far removed from the original classical theories invented by Newton.

But the standard model is not just a quantum theory. It is a quantum theory of a particular type, known as a quantum field theory, because it describes the forces via "fields," like the force fields of science fiction films—only these are said to permeate all of space and time.

Since general relativity is a classical theory, it and a quantum theory like the standard model are not compatible. To those who don't follow physics it may seem paradoxical that we have a classical theory of gravity and a quantum theory of the other forces, but the theories are generally applied to different situations, so the schizophrenia is manageable. Still, it is far from ideal, and many physicists today are trying to find a quantum version of general relativity, a theory of *quantum gravity*. The eventual goal is to construct a single quantum theory encompassing both quantum gravity and the standard model. That yet undiscovered theory would then describe all four forces of nature, which is why Einstein called it the unified field theory, and why physicists today call it the theory of everything.

When Stephen was starting graduate school, few physicists were interested in trying to develop a quantum theory of gravity, or a theory of everything. One reason is that, as I said, general relativity and quantum theories existed in a state of peaceful coexistence. They describe different forces, and they describe nature at different size scales. Just as, in biology, the study of mammals is separate from the study of bacteria, in physics the study of general relativity was separate from the study of quantum theories.

But Stephen was not like most physicists. Of the vast richness of ideas that is physics, what excited Stephen, once he entered graduate school and took his serious turn, was general relativity, in particular the subfield called cosmology, which seeks to use general relativity to understand the origin and development of the universe. Stephen was drawn to cosmology because only that field had the promise of answering the questions of existence that now most concerned him. To Stephen those who worked on the theory of elementary particles—which would eventually develop into the standard model—seemed less concerned with answering the deep questions of cosmology than with classifying the many particles and forces. What they were engaged in was like "botany," he said, and he wanted no part of it.

Hoyle, Stephen's first choice of advisor, was a big name in cosmology. He'd cofounded a theory of the universe called the steady state theory. Stephen was disappointed to have been assigned to Sciama instead. Once more in Stephen's life, what seemed like a setback turned out to be a blessing, for the more Stephen learned about Hoyle's steady state theory, the less he liked it. In fact, not long into his graduate student career, Stephen would make a splash by standing up during the Q&A after a research seminar Hoyle was giving at the Royal Society in London and announcing that he'd found a flaw in Hoyle's mathematics. A few years later, Stephen rubbed salt in the wound by making the first chapter of his Ph.D. dissertation a criticism of the steady state theory.

Hoyle was a great physicist who did pioneering research on how heavier elements are created from hydrogen and helium by the nuclear reactions in stars. But as a scientist he had a serious flaw—the inability to recognize that a pet theory, in this case the steady state theory, had been fatally undermined by the evidence. And so, had Hoyle been Stephen's advisor, those would not have been fun times. Sciama, on the other hand, was also a leading cosmologist, but he'd become disillusioned with Hoyle's theory, so Stephen's disdain for it caused no conflict.

Though Stephen was lucky in the advisor he was assigned, there was one problem with his choice of specialization: Stephen knew little about cosmology. He had studied physics as an Oxford undergrad, but he hadn't learned much. That he could advance as quickly as he did and within a few years make a name for himself in the research world tells you something about his brilliance. It also tells you something about physics.

In theoretical physics you can learn fast because the field depends on understanding concepts rather than absorbing a lot of facts, as in professions such as law or medicine. "You don't need to memorize," Stephen once told me with a grin. "You can derive." That's because physics condenses experience into a compact form. Einstein's equations, for example, can be written in just a line, but they encode the behavior and properties of countless systems, from the planetary orbits, to the flight of soccer balls, to the collapse of stars into black holes.

That Einstein's compact equations can be so powerful is no magic. The few symbols they are comprised of represent concepts that require enormous effort to fully grasp. We all, to an extent, condense our experience. The world would be too complex to understand if we didn't. We don't note that Fords stop at red lights, Toyotas stop at red lights, Volkswagens stop at red lights, and so on. We encompass all those observations in a single principle, or law: "Cars stop at red lights." That's just what we physicists do, only a thousand times over, and we write those laws in the elegant form of mathematics, which allows us to derive one law from another. Lawyers can't do that because, though there might exist general guiding principles, human laws are ad hoc creations and cannot be derived from each other. Nor can doctors derive the details of human anatomy from any set of first principles. That the laws of physics can is a wonder that every physicist marvels over.

Stephen studied those books and papers into which our knowledge of cosmological principles was condensed, and he learned fast. He expected to die in a few years, but at least in cosmology he would be spending his time addressing questions that excited him.

After they were married, on Bastille Day, 1965, Stephen and Jane rented a tiny old house at 11 Little St. Mary's Lane, near the medieval church in old Cambridge it was named for. The house had small rooms and low ceilings. It had recently been renovated, but there was no furniture and they had little money to buy any. The Hawkings bought a bed, a dining room table, some chairs, and a refrigerator about which they'd debated an entire afternoon.

Stephen was twenty-three and a graduate student. Jane was twenty-one and had one year left to complete her undergraduate degree in languages in London. She spent most of the week there, but they were together on the weekends. The advantage of the Little St. Mary's house was that it was just a hundred yards from where Stephen, at the time, had his office. But if getting to work was easy, living at home wasn't. The master bedroom was on the second floor, up an awkward spiral staircase. So was the bathroom. There was a third floor, too. This was when Stephen could still live alone and take care of himself. Still, to get to the bathroom he had to grab the rope banister and pull himself up the stairs like a rock climber. It took him a good ten minutes to do it, and even when others were around, he'd refuse help. "It's good exercise," he'd say.

Stephen did his best back then to ignore his illness, even when it was clear to everyone else that it was getting in the way. One time he set out to meet a friend, Robert Donovan, for dinner. He showed up with torn pants and scrapes on his face. He'd obviously had a bad fall on the way, and Robert was worried about him and wanted him to see a doctor. But Stephen didn't want to get patched up. He didn't even want to change clothes. He wanted to go eat as if nothing had happened. So that's what they did.

Before marrying Jane, Stephen had lived in a Cambridge residence hall that was really just a big old mansion divided up into student rooms. There was a large lawn and garden in back. Sometimes people played croquet out there. Stephen's room had a veranda that opened onto the croquet lawn. Very Victorian.

One day, Stephen's parents, Frank and Isobel, came over for afternoon tea. Both had attended Oxford. After graduating, Isobel took a job she didn't like and was extraordinarily over-qualified for—secretary in a medical research institute. But it was there that she'd met Frank, a doctor specializing in tropical medicine who had turned to research. Frank wanted Stephen to follow in his footsteps and become a doctor. Of Stephen's three siblings, only his sister Mary would do that, and Stephen never gave it serious consideration.

On the day of his parents' visit, Robert Donovan joined them. He was a chemistry graduate student one year behind Stephen. Stephen and Robert had met the day Robert arrived in Cambridge. Robert had been assigned to live in the same resi-dence but couldn't find anyone to unlock his room, so he started talking to this guy he saw practicing croquet shots by himself on the back lawn. He noticed that the fellow's gait was a bit off. That fellow was Stephen, and they would stay friends for fifty years.

After an hour or so of tea with Stephen's parents, Robert got up to leave. He had work to do. Frank followed him out. Frank had always been a remote father who disappeared each winter to spend months conducting research in Africa. But when Stephen took ill, Frank assumed a more active role. He'd even pleaded with Dennis Sciama to allow Stephen to cut corners in order to complete his Ph.D. dissertation early, before he died. Sciama turned him down.

Frank didn't know Robert, but he followed him out because he could tell that Robert and Stephen were good friends. "Please make sure Stephen is okay," Frank said. "Keep an eye on him. Let me know." Stephen came out after them. He was fuming. He was loud. "I can look after myself," he told his father. "I'll ask my friends myself if I need help. You don't do that."

Robert nodded, but gave Stephen's father a glance that said, *Don't worry, I'll do as you asked.* That was in 1963. Robert and Stephen grew very close, almost as close as Stephen and Jane would be. Robert and his wife named their first child Jane, and Stephen and Jane named theirs Robert. They would see each other often for the next seven years, until Robert left for a position in Edinburgh, where he would remain for the rest of his professional life.

Despite the distance between them, and despite the fame Stephen would achieve—and its demands, which have ruined many a friendship—Robert and Stephen remained best friends. Stephen enjoyed visiting Edinburgh, and Robert was always ready to come down to Cambridge for a family occasion. Or for one of Stephen's famous parties. Or just because Stephen would say he'd like to see him. Later, after many years had passed, after Stephen had lost his ability to be physically independent, Robert still looked back at that afternoon tea with fondness. It reminded him of the time when Stephen's body could still live up to his spirit.

Stephen's father died in 1986, his mother in 2013. In Robert's seven years at Cambridge, he never once felt he had to sound the alarm to Stephen's father. And in all their ensuing decades of friendship, Stephen never acted toward Robert as if there was anything to be alarmed about. Until one night in 2017. It was December, a cold dark month in the university town. Stephen had by then been divorced from Jane for over twenty years, but he still saw Robert often. They were getting ready to go out to a grand dinner at the college. Stephen started typing into his voice synthesizer.

"I don't think . . ." he said.

That was all that came out. Stephen kept typing. Robert waited. Five or six minutes passed. The words were taking forever to come out. Finally they did.

"I don't think I have long to go," Stephen said.

Robert was taken aback. Stephen didn't seem ill to him. Why was he saying this? As he had on the night of Stephen's bad fall decades earlier, Robert urged him to seek help. Or at least stay home and get some rest. But Stephen didn't want that. He wanted to go to dinner. They did, and that was the last time Robert saw him.

4

Those who have lost the capacity for most physical pleasures value those that remain even more. A touch. A symphony. A scent. A taste. Meals were always important to Stephen. But they were also a time of connection. They were a hiatus from the world of mathematics in favor of engagement in the human realm. Still, even at mealtime, there was no hiatus in Stephen's keen thought processes. If his approach to physics had the same mischievousness as his approach to humanity, his approach to humanity had the same sharpness of intellect that he employed in his physics.

Once at dinner when Stephen was still working on his Ph.D., he found himself at a table at Trinity College with a South African engineer. The engineer had just arrived in Cambridge. But if he marveled, as I had, at the university, and especially at Trinity—where Newton had studied—he kept that to himself. What he did do was proudly hold forth about how well South Africa was faring. According to the engineer, these were fabulous times in his homeland.

Stephen was never one to hold back on what he thought. On one occasion, after he'd become famous, he'd been invited as an honored guest to watch a modern production of the opera *Madame Butterfly* in Berlin. It turned out to be a mediocre performance. Afterward, the principal, delighted to host Stephen, said, "Professor Hawking, what did you think of the performance?" Stephen replied, "It wasn't very good, was it?" His host was surprised at that answer. But then he said to Stephen, "Yes. I agree."

The South African engineer's disquisition couldn't be described as either mediocre or not mediocre. It was just his opinion, expressed at too great a length. But it got Stephen's attention. Stephen had ideas about it, and he wasn't going to keep them to himself. He asked the engineer, "What about the black people?"

"They don't really count," the engineer said. In the early 1960s, that was not an unusual reply.

"Why don't they count?" Stephen asked.

"They can't take care of themselves," he said. He talked about apartheid. It works, he said, and it's necessary.

Although Stephen didn't argue, he kept asking questions. He was challenging the man's beliefs not by disagreeing with them, but in the Socratic style, by forcing him to see the implications, unadorned, of what he was saying.

The engineer had started the conversation, stating what he "knew" to be true. He had never before examined it very closely. But Stephen made their dinner conversation an exploration of the man's beliefs, an exploration that the engineer himself had obviously never made. In the end, the man was flustered. Now that he'd been pushed to understand the foundation of his beliefs about apartheid and the nature of black people, he was questioning himself.

I once had a physics professor who advised me that "if you like asking questions and searching for answers, become a physicist. If you like learning the answers and applying them, become an engineer." That's a broad generalization, but it illustrates a difference in both the philosophy and psychology of those fields. Are you one who tends to learn and apply knowledge, or to question and create it? In leading the engineer to question, Stephen was just being Stephen, for it is by questioning your beliefs, and those of others, that important discoveries occur, not just in life but in physics.

The engineer had viewed his country as most people view the night sky, as constellations of white dots floating in a vast and unimportant sea of black. Through his questioning, Stephen led him to see more than just the dots. Stephen would do that to his fellow physicists, too. While they marveled at the stars and the galaxies, Stephen raised questions about the *space* in between them. *Where did it come from? How did it all begin?* In trying to understand the meaning of our existence, those seemed to him to be the most basic of questions. Yet, as Stephen began his Ph.D., few were asking them.

This was during the era in which general relativity and cosmology were in the doldrums. Physicists' lack of interest in the birth of the universe seemed to make sense, because physics is an empirical science while the origin of the universe is not something we can observe directly. Because light takes time to get to us, by observing the light from distant galaxies we can essentially look back in time, but not that far back. Nor did anyone, in the early 1960s, know of any indirect way to test a theory of the universe's origin. As a result of such issues, physicists tended to consider cosmology a pseudoscience, a mathematical playground outside the realm of experimental testing. That would begin to change after the accidental discovery in 1964 of the faint afterglow left over from the big bang, called the cosmic microwave background radiation. When Stephen was starting out at Cambridge, that was still a year or two away.

Another issue back then was the difficulty of understanding just what Einstein's theory actually does predict. Like any theory in physics, Einstein's is a scheme of mathematics and rules about what it represents and how to apply it. To extract what the theory has to say about a particular system, you have to use the scheme to set up equations tailored to that system and solve them, or at least approximate the solution. Einstein's equations are in most cases extremely difficult to solve, so today we investigate their implications through the use of supercomputers, but the computer power available back then was feeble in comparison.

Due to difficulties of that sort, when Stephen moved to Cambridge the practitioners of general relativity and cosmology were mainly mathematicians whose work was detached from reality and whose models of the universe were unrealistic. That kept them occupied, but nobody paid much attention to their papers. The low quality of the work prompted Caltech physicist Richard Feynman to write his wife from a 1962 conference on gravity in Warsaw, "Because there are no experiments this field is not an active one . . . there are hosts of dopes here and it is not good for my blood pressure: such inane things are said and seriously discussed that I get into arguments . . ."

Most physicists concurred that questions of the origin of the universe were dead ends, but those were the questions that had won Stephen's heart. And so, rather than being discouraged, he viewed the fact that the field was a backwater as an advantage. To him, the field wasn't "dead," it was "ripe"—and he was just the person to pluck it.

To nonscientists it may seem that theoretical physicists mainly solve problems. But what is more important than solving problems is posing them, because the questions you ask govern the answers you find. Questions are both a reflection of, and a determinant of, the way you look at the world. Stephen had a knack for dismissing what would later turn out to be unimportant and for quickly identifying what was really the heart of a matter. He intuitively asked the right questions and questioned the dubious assumptions of others. Because of that, Stephen was viewed as a rebel. That role came naturally to him—he ignored conventional wisdom just as he ignored the speed limits and the advice of his doctors. He drove his car wildly and recklessly, and his physics was also wild and unrestrained. But it wasn't reckless. In physics, Stephen always knew, even as a graduate student, where he wanted to go, and why.

Physics is supposed to be a field of reason and logic. That is an important part of it, but in order to reason logically, you must first have a framework of thought that defines the assumptions you are making, the concepts you will use, and the questions you seek to answer. People often accept the frameworks they inherit from others, from history, or from their past, and never question or sufficiently examine them.

With regard to Stephen's burning *How did it all begin?* question, for two millennia everyone had assumed that the universe had either always been in existence and was unchanging, or else that it was created at some moment—for example, as described in the Bible—and had been relatively unchanging since then.*
Philosophers from Aristotle to Kant, as well as scientists, including even Isaac Newton, believed this.

Newton should have known better. How could a collection of galaxies and stars maintain a stationary configuration when each, through the force of gravity, pulls all the others toward it?

---

* By "unchanging" they meant on the cosmic scale. Obviously, small-scale change is part of nature—planets orbit, rocks fall, people live and die.

Shouldn't the objects coalesce over time? And, since forever is a long time, shouldn't all matter, by now, be clumped together in a big dense ball? Newton was aware of this issue but talked himself out of it by convincing himself that if the universe were infinitely large, the clumping wouldn't happen. That is wrong. Others, after Newton, tried to modify his theory to make gravity repulsive at long distances, employing a mathematical alteration small enough that it wouldn't noticeably affect the orbit of the planets but large enough that it keeps the universe from collapsing upon itself. They were unsuccessful. Even Einstein joined the game. He added an extra "anti-gravity" term, called the cosmological constant, to the equations of general relativity to supply the repulsive force needed to keep the cosmos from contracting.*

The realization that all these eminent philosophers and scientists were misguided—that the universe is changing, expanding, evolving—is one of the most extraordinary discoveries of the twentieth century. It is due to American astronomer Edwin Hubble, who had taught Spanish and coached basketball in a high school in New Albany, Indiana, before deciding to pursue a Ph.D. at the University of Chicago.

After graduating, Hubble had the good fortune to arrive at the Mount Wilson Observatory near Caltech in 1919, just as a new telescope was being installed. At the time, the prevailing view was that the universe consisted only of the Milky Way galaxy. Then, in 1924, Hubble discovered that the specks astronomers see when they examine nebulae—whitish clouds that stretch between the stars—were made of other, distant galaxies.

---

* The cosmological constant acts only on very large scales. It introduced no effects that could be measured with the technology available at the time, and so whether or not to include it was a choice Einstein was free to make. In 1998 that changed. The term really does need to be there.

Those clouds of galaxies seemed to extend as far as the Mount Wilson telescope would allow him to see. We now know that they extend even farther.

Because stars are hot, the atoms in their atmosphere are in a state of high energy. Some of that is energy of motion, but some is stored internally, in the electrons within the atom. Quantum theory tells us that the energy of those orbiting electrons can only take on certain values. When an electron leaps from one such energy level to a lower one, the atom emits light with a frequency that reflects the energy difference between the level the electron came from and the one it landed in. But each element has a unique set of energy levels. As a result, atoms of hydrogen, helium, and other elements each emit light that consists of a unique set of frequencies. That light provides a fingerprint that can be used to identify the element it came from. Astronomers use that fingerprint to identify the composition of comets, nebulae, and various types of stars.

In his years at Mount Wilson, Hubble noticed that compared to the light we observe coming from atoms here on earth, the light emanating from other galaxies was shifted toward the lower frequencies, the red end of the spectrum. He also noted that the farther away the galaxy, the greater that "redshift."

The frequency shifts that captured Hubble's imagination are due to a phenomenon first studied by Austrian physicist Christian Doppler in 1842. Doppler found that the color of light you observe from any given source depends on its motion with respect to you. The light will be redder if its source is moving away from you, and bluer if it moves toward you. Taking Doppler's theory into account, Hubble's work showed that the galaxies were moving away from us—and the more distant they were, the faster they were moving. This led to the dizzying conclusion that the universe is not only far vaster than anyone had thought, but that it is also expanding.

Astrophysicists sometimes use a "raisin bread" analogy to explain this thinking. Before I describe that, it is important to understand that the expansion of the universe is not like the explosion of a bomb. A bomb blows hot gases and shrapnel out into space that is already there. But there is no "outside" of the universe. When physicists say the universe is expanding, we mean that space itself is growing, from within: if you pick any two points the distance between them increases over time.

The raisin bread analogy works like this: Imagine you are inside a ball of dough riddled with raisins that are roughly evenly distributed. The ball of dough represents three-dimensional space. The raisins represent the galaxy clusters. This model has a flaw, in that space therefore has an edge, the outer surface of the dough ball. Space has no such edge, but for the purpose of this analogy that's not an important distinction. Now let the dough rise until its radius has doubled. A raisin an inch away from you will then be two inches away—an inch farther than when we started. In such a scenario a raisin that started three inches away will now be six inches away. It will have moved three inches in the same amount of time, so the speed at which it is moving away from you will be three times that of the first. Similarly, a raisin that started five inches away will now be at a distance of ten inches, having moved five inches in that time. As the dough keeps expanding all the raisins will move away from you, and the farther away a raisin, the faster it will recede.

In 1929, almost a century after Darwin started to formulate his theory of biological evolution, Hubble had discovered that the universe, too, was evolving. But the idea of an unchanging universe did not die easily. Physicists, who are good at such things, concocted theories to save their precious preconception. One of the most famous came from Fred Hoyle's work on the steady state theory. Adherents of that theory didn't dispute that the distant galaxies were moving away from us, but the theory postulated that new matter is constantly being created, so that the density of matter remains unchanged as the universe expands, with the new matter filling in the new space. In that way, the universe could remain, on the cosmic scale, unchanged.

The steady state theory's main competitor at the time was the big bang theory. Hoyle wanted nothing to do with that latter theory, but he was responsible for its name. It came from a comment he made on a BBC radio broadcast in 1949, when he called it "the hypothesis that all the matter in the universe was created in one big bang at a particular time in the remote past." Some say he'd used the term sarcastically. He denied that. Either way, the term stuck.

If a theory draws a lot of interest, one of the first things physicists will do is name it. A measure of the lack of interest in the big bang theory was that it didn't get its name until about twenty years after it was conceived. The theory had been invented by a brilliant Belgian priest and physics professor named Georges Lemaître. He began by analyzing Einstein's equations, which in 1927 led him to argue that the universe must be expanding—two years before Hubble's work showed that this was indeed the case. He then noted that if the universe is growing larger, it must have been smaller in the past, and that the farther back you go, the smaller the universe. He concluded in 1931 that at some time in the past the size of the universe must have been zero—in other words, that all the mass of the universe must have been concentrated into a single point. He called that the "primeval atom."

The big bang theory seemed to imply that there was a moment of creation, but again, clever physicists found a way to avoid that conclusion. They created a version of the big bang theory in which, going back in time, matter didn't all contract to a single point but rather to a small volume, so that, as time moves backward, particles of matter could skate past each other. As a result, instead of crashing into a single point, particles would come close, fly by one another, and again move farther apart. In that way, the universe would be eternal, but it would exist in alternate cycles of expansion and contraction. Belief in the steady state and the various forms of the big bang was divided when Stephen entered Cambridge—at least among those few physicists who gave it any thought at all.

Stephen told me once when I brought up the topic of religion that he did not "engage in metaphysics." Like the philosophers, Stephen wanted to answer the great questions, but he wanted to do it using science, which he knew was much harder. In science, reason is not enough. In philosophy you are free to theorize. In science, experiment can prove that you are wrong. Stephen felt that scientists from Newton to Einstein had been let down by their philosophical and religious beliefs, seduced into ideas about physics that were not backed up by theory or experiment. And so he questioned the belief that the universe is unchanging and that it is eternal. Just as important, he questioned the even more widespread belief that the issue is not one of great importance.

In the repository at Cambridge, stamped with the date 1 FEB 1966, is Stephen Hawking's Ph.D. dissertation: *Properties of Expanding Universes*. He was twenty-four then. The dissertation opens, "Some implications and consequences of the expanding universe are examined . . ." Typed by Jane—because Stephen was incapable of doing it—what follows are four chapters that include cross-outs and handwritten equations. The last chapter, about twenty pages long, is the one that made Stephen famous among his peers.

Stephen had arrived in Cambridge in October 1962. He spent his first two years of graduate school making lifelong friends and settling into married life, but in his physics he was drifting. He was studying general relativity, attacking various problems that he and his advisor, Sciama, thought promising, but not discovering anything much of note.

Those studies, which would form the first three chapters of his Ph.D. dissertation, were unremarkable. They were independent mathematical analyses of various topics, and they made interesting points, mainly mathematical criticisms of Hoyle's steady state theory. But the work had holes and left unanswered questions. Alone, those chapters might not have been sufficient to earn Stephen a Ph.D., and they certainly would not have made him famous. But thanks to Stephen's becoming familiar with the work of a thirty-three-year-old mathematician named Roger Penrose, he added a fourth chapter, more or less independent of the others, and this was the chapter that would launch his career. Stephen learned of Penrose's work in January 1965, after Penrose gave a seminar on it at King's College, in London. Ten years younger than Penrose, Stephen had been attending that series of seminars. As it happens, he didn't attend this one, but he heard about it from Brandon Carter, his office mate in Cambridge.

If, in the story of the universe, it is important to take into account the pull of all matter toward all other matter, it is equally important when telling the tale of a star. One might wonder, for example, why the sum of all those attractions doesn't cause the star to collapse upon itself. The answer comes from the nuclear reactions within the star. They lend the stars heat, imparting to the gases a tendency to expand, thus balancing the compression effect of gravity. The work Penrose described in his talk concerned what happens after that, after a massive star burns out its nuclear fuel and begins to cool down. When that happens, the dying star begins to collapse under the force of its own gravity.

Penrose recognized that the collapse is a complex and chaotic process, and doesn't necessarily maintain the original star's neat spherical symmetry. As a result, the collapse can proceed according to two possible scenarios. One is reminiscent of the version of the big bang theory in which the particles skate past each other: as the star collapses, its constituents could all fall toward the star's center but not toward precisely the same point. They might then race past one another, resulting in an expansion phase. In the other scenario, despite the chaos of the collapse the stuff of the star is all drawn to its precise center, where it is crushed into a single point in which the density of matter is infinite.

That second possibility, Penrose eventually proved, is the one that Einstein's equations demand. In 1969, physicist John Wheeler would call dead stars of that sort—those with a point of infinite density at their center—black holes, but in 1965 there hadn't yet been enough interest in them to produce an agreed-upon name.

Physicists call a point at which physical quantities become infinite a singularity. Physicists shun singularities because we shun infinities. We shun infinities because, though they may occur in mathematics, they don't occur in the real world. Nothing we measure is infinite, so any theory that predicts that a singularity occurs must be wrong.

As a workaround, physicists tried to find a way to render the presence of the singularity moot. They thought of a few options. One is to point out that Einstein's theory is not a quantum theory, and therefore at some point during the star's collapse—when it reached a certain minute size—his theory would no longer apply without some (yet to be invented) modification. Will that modification eliminate the singularity? We don't know. Another is to say that since we cannot look inside a black hole, the singularity is forever hidden—unobservable—

and hence does not matter. That sounds reasonable, but it's not that simple. Black holes can rotate, and some rather esoteric calculations suggest that that rotation might expose the singularity. So the jury on this is still out.

The famous chapter that Stephen added to his dissertation didn't have to do with any of that. While Penrose's work had stimulated many theorists to start pondering black holes, Stephen, as usual, took off in his own direction. He saw the story of the star collapsing under the force of gravity as being reminiscent of the big bang, only in reverse. What if the universe was like a giant black hole that, if you run time backward, collapses just like one of Penrose's stars? Could he adapt Penrose's mathematical methods to gain insight that escaped even Einstein? Could he prove that Einstein's equations dictate that the big bang—and not the version that includes repeated cycles of expansion and contraction—must have occurred?

Like Galileo, who had taken a primitive spyglass, improved its optics, and directed it toward the heavens, Stephen appropriated Penrose's mathematics and employed it to study the cosmos. With chapter four of his Ph.D. dissertation—and follow-up work done together with Penrose himself in the years to come—Stephen soon surpassed his advisor Dennis Sciama in reputation and eventually even his once-desired advisor, Fred Hoyle: he showed that, singularity and all, the big bang is an unavoidable consequence of general relativity. There were no cycles of expansion and contraction, there was a beginning, and at that moment, though physicists didn't like it, the universe was packed into a space of zero volume. At least, those conclusions were the inevitable result of Einstein's equations.

Around the time Stephen was doing his theoretical work, observational astrophysicists started to find experimental evidence of the big bang. Nuclear physics had shown that, in the first minutes after that event, the extremes of temperature and pressure would cause some hydrogen nuclei (protons) to fuse together, forming helium. Detailed calculations had indicated that we ought to find about one helium nucleus for every ten hydrogen nuclei, and astronomical observation confirmed this. The big bang theory also predicted that some radiation from that event should persist to this day—the cosmic microwave background radiation. That, too, had been discovered, two years before Stephen's dissertation. But the mathematics proving that the big bang is a necessity of Einstein's equations—that came from Stephen, in his first major foray into the world of physics.

5

Several months had passed since my punting visit. I was in Cambridge again. We'd been working since I arrived a few days earlier, making slow progress. On this day, Stephen had sent me an unusual morning email. It contained a suggestion that threw me off a bit, so I was anxious to discuss it with him. Until now, we'd been in agreement about what we wanted to say in the book, but in the email he seemed to be doing an about-face on an important topic.

As I turned off the staircase and stepped toward Stephen's office, I could see that his door was closed. By now I knew what that meant, so I decided to hover in the hallway for a while. I looked over the green chalkboard to the left of his door. It was an anachronism. The building was generally modern: Stephen's black door; its metal lever-style knob; the purple wall and bright yellow bulletin board with announcements of upcoming conferences. But that dusty chalkboard, in this whiteboard/dry-erase era, was a throwback. So were the diagrams students had scrawled upon it, so-called space-time diagrams that help physicists envision processes in general relativity. They were antiques, a type of diagram invented in 1907 by one of Einstein's former professors, Hermann Minkowski.

I thought about Minkowski. A century ago, working in Zurich, he had had a "big idea" and scrawled it on his own chalkboard. The idea, inspired by Einstein's special relativity, was to include time on an equal footing with the three directions of space in the mathematics of the theory. Einstein's was a momentous breakthrough, but it was Minkowski who gave the concept of space-time the formal meaning we assign it today.

We all like to speak of big ideas. In physics, at least, they are not an end but a beginning. When you have an idea, any idea, one of the challenges in physics is to work out the implications and mathematical details that connect it to our greater body of knowledge and make it meaningful. In the case of space-time that meant, among other things, defining what is meant by "distance," now that time is to be considered a fourth coordinate. We're all familiar with the idea of distance between two places, but what is the distance between A and B when those points each represent both a place and a time?* Minkowski's mathematical solution to this need not concern us here. What is important is that he did find an answer, and that his new concept of distance was an important reason his notion of space-time had an impact. It would, in fact, play a key role in Einstein's later development of general relativity.

Said Minkowski, upon presenting his findings, "The views of space and time which I wish to lay before you . . . are radical. Henceforth space by itself, and time by itself, are doomed to

---

* Space is made of points that can be specified by latitude, longitude, and height. You determine how far one point is from another through the difference in these coordinates. Space-time is built from "events," the points of space with a time stamp appended, and the separation of events in that space depends on their time difference as well as their spatial distance.

fade away into mere shadows, and only a kind of union between the two will preserve an independent reality." His prediction came true.

Standing in that hallway, it struck me that each time we think of Minkowski's space-time we are transcending it, connecting with him on the spaceless and timeless plane of ideas. I felt goose bumps as I realized that Stephen wielded an influence of that same magnitude, that some wide-eyed physicist in awe of his imagination would stand before some diagram and equations a century hence and feel a similar connection with Stephen.

Like Minkowski, Stephen had taken relativity a leap beyond where he had found it. But Stephen's leap was in a direction Einstein might not have approved of. Einstein had not been fond of quantum theory, and his general relativity violated its principles. In the decades following his development of general relativity, in which few worked on that theory, the incompatibility did not bother many physicists. But by finding ways to employ general relativity and quantum theory in the same realm—in his theories about the early universe and about black holes—Stephen would demonstrate the potential of that combination, and lead the physics of relativity in a new direction.

General relativity and quantum theory—two gifts of the human intellect, beautiful and remarkably successful. Both have shaped today's technology, and both have shaped physicists' understanding of nature. And yet they can't both be right. They clash; they contradict each other. As I got to know Stephen better and to understand his character, I realized that reconciling contradictory theories and ideas was one of his great strengths. It came as naturally to him as does migration to a bird. After all, he was a man both dead and alive, both powerful and powerless, both daring and careful. With Stephen, contradiction was not just a philosophy of life, it was a way of life.

As I waited for Stephen's door to open I thought about the vast amount of material Stephen and I needed to go over and how our time together was creeping along, like the trickle of words Stephen managed to utter. The communication bottleneck was something you had to get used to—the sitting, waiting for him to compose his words and sentences.

In the first two decades after his diagnosis, Stephen's speech had gradually deteriorated. Eventually, only a few could understand him—Jane, Kip, Robert Donovan, a few of his Ph.D. students. They would act as his translators, and he couldn't communicate except in their presence. But in 1985 Stephen had a severe lung infection. He was forty-three years old. He spent weeks on a ventilator, and each time his doctors tried to wean him off it, he fell into a choking fit. His doctors told Jane that Stephen's only chance for survival would be a tracheostomy, and they explained that it would be irreversible, meaning that he could never speak again. Stephen was deemed too ill to approve the operation, so the decision was hers. She signed off on it. Stephen recovered, but after the operation the only way he could communicate was through the use of a spelling card. A companion would point to various letters on the card, and Stephen would raise his eyebrows when they'd selected the one he needed.

Stephen was alive but felt dead. He found it difficult to accept that the operation had really been necessary. He was furious at Jane for agreeing to it. That period, in which he could not communicate, was the only time after his initial diagnosis when Stephen let his illness get him down. He fell into a deep depression.

After about a year, Judy Fella, who was then Stephen's assistant, saw a report about a computer program for the severely disabled on the BBC. She tracked down the inventor, and before long Stephen had been outfitted with the initial version of the communication system he would use for the rest of his life. With the new technology, composing a sentence became, for him, like playing a video game. The cursor would move on the screen, and he would capture the letter or word he wanted by moving his cheek to activate a sensor in his glasses. When he was done, he would click on an icon and his famous computer voice would read what he had typed out. When he was on a roll he got out about six words per minute. It wasn't fast, but at least he could communicate. What's more, he no longer needed a translator. That meant that for the first time in many years, he could have a private conversation with anyone he pleased.

I'd had some practice at the six-word-per-minute waiting game when we'd met at Caltech while collaborating on *Briefer History,* and while creating the plan for *Grand Design.* But I still wasn't accustomed to it. Sometimes the wait for a response was a minute, sometimes five or ten. At first my mind would wander. Then I learned to relax, to put myself in a semi-meditative state. But writing *Briefer History* was a molehill and writing *Grand Design* a mountain. I couldn't meditate myself up that one.

Working on *Grand Design,* I learned to use Stephen's typing time to ponder the issue at hand. I learned that the slowdown in our give-and-take was valuable. I could think more deeply and consider the issues more intelligently than one does in ordinary conversation, in which all parties answer immediately— which means off the top of their heads. Sometimes I mused that everyone should converse in this manner. Other times I thought no one should be made to walk through this field of molasses.

As we grew closer, our interactions evolved. I'd learn that composing words was not the most important way Stephen communicated. Just as the blind develop an increased power of hearing, Stephen had developed an amplified capacity for nonverbal communication. His close friends learned to take advantage of that, to speak in a provocative or leading manner and then observe his reaction. By watching him closely as they spoke, his friends would probe his mind indirectly, employing their own words, like physicists studying an atom by observing how light scatters off it. Stephen would interject a word or sentence when necessary, but his most powerful means of transmitting feelings was by facial expression, the nuances of which he could manage with his eyes, brow, and mouth. Some of his expressions, like his grimaces, were obvious, others were subtle. Sometimes you'd feel you understood what he meant, and not be quite sure how he'd gotten it across. It was a special language, and when you got close enough to Stephen, you'd learn it, just as, before his tracheostomy, those who were close to him learned to understand his garbled utterances. For Stephen, voiced words became the spice of your conversations, not the meat.

Stephen's door was still shut. It had remained shut for quite a while, and I was growing tired of the waiting. I looked over to Judith in her office, looking through a stack of mail while speaking at a fast pace into the phone, which she cradled on her shoulder.

Stephen's world was a frenetic one. And if things seemed to settle down a bit, he'd find a way to kick it up a notch. He tried to kick everything up a notch, including *Grand Design*. When, after we'd written *A Briefer History,* I asked him if he wanted to write another book together, my idea was to focus on his latest physics research, which I had been following with interest. Had we stuck to the physics alone, that would have been plenty for a great and provocative book. But Stephen soon made the idea bigger.

In *The Grand Design* he wanted to talk about the *philosophical* implications of his latest work. "I want to present a new philosophy for theoretical physics," he told me. That was a rather bold goal. It interested me as long as we didn't take it too seriously— after all, we weren't philosophers. We weren't experts in the field. On the other hand, I didn't see anything wrong with trying to illuminate how physicists think about their work and its relation to the world, and the science content of the book did invite such a discussion.

Given Stephen's desire to get philosophical, I was surprised when I saw the message that he had emailed me that morning. It included some text that he suggested be added early in the first chapter. His text began, "How can we understand the world in which we find ourselves? How does the universe behave? What is the nature of reality? Where did all this come from? Did the universe need a creator?"

And then he added, "Traditionally these are questions for Philosophy, but Philosophy is dead . . ."

How was it possible, I wondered, to start a book in which we'd present "a new philosophy for theoretical physics" with the statement that philosophy is dead?

I was contemplating that when Judith finally hung up. She smiled and blasted out my name: "Leonard! Good morning!" I said good morning back, though it was the afternoon. I'd learned to arrive sometime between noon and 1:00 p.m.

She waved me into her office. It was crammed with books and papers, and boxes containing more books and papers. On some shelves were foreign versions of *A Brief History of Time,* many in languages I didn't recognize. As Stephen once said, he knew *Brief History* was doing well when it came out in Serbo-Croatian.

Judith was sympathetic to my impatience. "You must be a

saint to work with him," she said. And then, "Look at this stack of mail I have to go through! Every day it's like this. You must look at *this* letter. You'll appreciate it! It's amazing!" She handed me a note addressed to "Professor Hawking," two pages in pristine handwriting with plenty of flourishes. Its author began by saying, "Words cannot describe the pleasure I obtain from your work. You may recall my previous good wishes, accompanied with handmade truffles sent from London . . ." And then a few lines down, "I had a LOVE and HATE relationship with physics. On December 25, 2005, I met Jesus Christ at the doorstep of my then London residence . . . Jesus was on crutches. He was a blond, young man with Ox-bridge [Oxford/Cambridge] seal on His forehead. He gave me the assurance telepathically that I would be amazed, at realizing how simple His system is/was, and I would soon be able to learn his technics in running the universes, or at least our universe . . ."

I asked Judith if she showed such letters to Stephen. "No," she said. "He gets annoyed when I take his time with them. He'll switch off. People send their theories, or letters about aliens. He thinks they're crackpots. But I'll answer for him. I love crackpots. After all, the woman who wrote this is interested in exactly the same thing Stephen is. She wants to understand the system of the universe."

True, I thought. And what a wonderful universe it would be if, instead of having to assimilate thick books of complex mathematics, I could have it explained to me by a telepathic Jesus on crutches. How wonderful if, considering the source of that telepathy, I could trust that the theory being conveyed to me actually worked. There'd be no need for experiment, no danger of what you believe being debunked. I found myself envying the woman who wrote the letter. The insane have it good, I thought, at least in this respect.

And then I thought, I *do* have what she has, or at least a

variation on it—I have Stephen. Stephen amazed me with his view of the system of the universe, and he communicated it magically, if not telepathically. He even fit the Oxbridge label, except that it is not imprinted on his forehead. The biggest difference is, sadly, that Stephen's theories, not being of divine origin, were not guaranteed to be correct.

Stephen's door finally opened. He was ready for me. I was ready for him, too. I had been for some time. But he didn't acknowledge me as I walked in. He was being fed spoonfuls of tea, and vitamin pills. His carer would dip that large pilfered tablespoon into his cup, add a few pills, and hold it to his mouth. He would open up enthusiastically, and she would move the spoon in. He was chronically thirsty, but more than that, he was always anxious to have his pills. A little too anxious, I thought.

Stephen took about eighty vitamin pills a day. He took pills every couple of hours, most of them given through his "peg," an opening in his belly through which his carers could inject fluids directly into his stomach. Early on, Stephen's father had suggested that a regimen of antioxidants like folic acid might help him. It was just a shot in the dark—and fifty years later there was still no evidence that such supplements were of any use. I thought at first that Stephen didn't really put much faith in the regimen, that he figured they couldn't hurt, so why not hedge his bets? I remembered the story that physicist George Gamow told about Niels Bohr, one of the founders of quantum theory. He said that Bohr had a horseshoe nailed above the front door of his country cottage in Tisvilde, Denmark. A visitor asked, "Being as a great scientist as you are, do you really believe that a horseshoe above the entrance brings luck?" "No," Bohr answered. "I don't believe it. But they say that it works even if you don't believe!"

It wasn't long before I discovered I was wrong. Stephen did not have Bohr's tra-la-la attitude. He felt very strongly about his pills. He had *faith* in his pills. So much faith that he was psychologically dependent on them. He was addicted, his carers told me.

Once, Stephen was at a conference in Texas when the Icelandic volcano Eyjafjallajökull erupted. Air traffic across northern Europe was interrupted for six days. Stephen was stuck and ran out of pills. He panicked and grew desperate. There was talk of asking his acquaintance Prince Philip of Spain (now the king) to lend him his private jet, either to bring Stephen his pills or to fly him back to Europe, but Judith didn't act on the idea. She let it die. "Who charters a plane to deliver vitamins?" she said.

Stephen was one of the most physically vulnerable humans on earth. He was unable to feed or care for himself in any way. He was brittle, frail, subject to chronic chest infection, weaker with each passing year. Despite all that he loved to socialize, to go to parties, and to travel the world. He was an adventurer. He went up in the "vomit comet," a specially modified Boeing 727 that takes off on the space shuttle's runway and then dives repeatedly to give thrill seekers experiences of zero gravity. He even had hopes of accepting an invitation from Richard Branson to launch him into space. Yet there was one circumstance that truly scared him—running out of his vitamin pills.

To Stephen's doctors, his belief that nutritional supplements were keeping him alive probably seemed no more likely to be true than those odd theories of the universe that Stephen got in the mail. He dismissed those theories, yet he shared with their authors the fundamental human need for understanding one's place and for wanting to exercise power over one's plight. The pills were a way to fight the disease, his father had said. In physics, Stephen could use mathematics to test such ideas. With regard to his illness, that was not possible, and so he clung to the recommendation of his loving physician parent and embraced the prescription that was his legacy.

Given Stephen's natural skepticism, his belief in the pills seemed out of character. Not that he wasn't open-minded. He was willing to consider, at least provisionally, any theory that didn't contradict known facts. And he wasn't bothered when different theories conceptualized the world in vastly different ways, as is true of general relativity and quantum theory. He was willing to accept them both, and to jump between their different descriptions as needed. The main quality that Stephen demanded of a theory was that it make predictions that could be proved or disproved through observation or experiment. "Any picture of reality is valid," Stephen told me, "if it agrees with observation."

Plato believed that we can know the world of mathematics, with its clear and unchanging laws, but that we can never have true knowledge of the physical world that we perceive with our senses. Stephen seemed to agree with Plato and to take his views a step further. Like Kant, he recognized that both our physical perception of the universe, and the concepts that form our mathematical descriptions of it, are shaped by our brains' architecture. The nature of the brain, in this view, determines how we think and the kinds of ideas we can create. Stephen believed that, as a result, scientists are compelled to view nature only in a particular manner, and are restricted to being able to comprehend only a limited range of theories. So the world our scientific theories describe exists only in our minds, and it is pointless to speculate about the existence of an "objective" reality.

One of Stephen's favorite passages in *The Grand Design* would be a story I dug up on the Internet about a town in Italy that barred pet owners from keeping goldfish in bowls with curved walls. Animal rights activists apparently considered that cruel, because the world outside the bowl would appear distorted. It's the kind of tale that would ordinarily make Stephen smile and roll his eyes, but the goldfish scenario illustrated an important point he wanted to make about our knowledge of the physical world.

Newton taught us that freely moving objects travel in straight lines. However, light bends as it passes from air to water. As a result, objects moving along straight lines outside the fishbowl would appear to the fish to be moving on curved paths. Now imagine a fish scientist formulating laws of motion to describe the motion of objects in that "outer space" beyond the walls of the bowl. One would expect those laws to reflect the fish's experience, and thus to dictate that free objects travel along curved paths. Though that theory would look odd to us, the fish could use it to make accurate predictions about the motion of outside objects.

Suppose that a particularly clever fish creates a new theory. This theory asserts that freely moving objects outside the bowl travel in straight lines. Those lines only *appear* curved, this fish postulates, because light bends as it passes from the outside to the inside world. The second fish's theory describes the same observations as the earlier theory, but it does it in different terms. While the first theory says that the objects move along curved paths, the second states that the paths are straight and it is the light that curves.

Though the predictions of the two theories are identical, some fish scientists would prefer the second theory, others the first. Or they might use them both, depending upon which is more convenient in a given context. There may also be some fish philosophers that would debate the question of which theory, if any, corresponds to "reality."

As humans reading this story we are predisposed toward the

second theory. That's because we have the "outside view." We are like gods to the fish, in that we created their universe and have experience in the outer realm, which they can never know. But from the point of view of the fish, which can't penetrate the glass bowl, the question of which theory best describes the outer world can never be settled.

Stephen believed that we find ourselves in the same situation as the fish. First, because the makeup of the human mind is a metaphorical bowl that limits the way we can understand the world. And secondly, because in modern physics we are increasingly presenting theories that, like the two fish theories, paint different and sometimes seemingly contradictory pictures of what is transpiring, while nevertheless agreeing in all their testable predictions. The famous particle versus wave duality was one of the first, but there are examples that go back at least as far as Copernicus.

In the second century A.D., Ptolemy created a model of the heavens with the earth at the center, surrounded by eight spheres that carried the moon, the sun, the stars, and the planets, which themselves moved on smaller circles—epicycles—attached to their spheres. In that way, the model could account for the complicated paths we observed in the sky. In the sixteenth century, however, Nicolaus Copernicus proposed the heliocentric theory that is now familiar to us, with the sun at the center. It is often said that Copernicus was correct and Ptolemy was wrong, that Copernicus taught us that the sun is "really" at the center of the solar system, and not the earth. But one can build an earth-centered model that describes what we see in the night sky as well as any heliocentric one. From the point of view of modern physics, the two approaches are both valid. The heliocentric model is superior in the sense that it is simpler, but that is a matter of practicality, or aesthetics.

Stephen believed that since the question of what is real

can never be settled, one should not spend any time on it. His response to such questions was to grimace, as if he'd just been fed a piece of rotten meat. To the extent that a theory makes predictions about the world that we can check and confirm, we should believe it. If an alternate theory describes the world differently but also makes predictions that check out, we should believe it, too, and use whatever picture of reality best suits our current purpose. And if an otherwise successful theory makes additional predictions about other universes or dimensions that are beyond our access, that is fine as well, and we shouldn't worry about whether those realms "really exist." That is Stephen's philosophy of physics.

Though Stephen had declared philosophy dead, by discussing such issues he was definitely flirting with it. His ideas fit squarely into a philosophical argument that has a long tradition: scientific realism vs. anti-realism. According to realism, the job of scientific theories is to describe the world accurately. According to anti-realism, the job of theories is merely to organize our sensory experience. Stephen's ideas about reality seemed to be a hybrid. Because of that, I thought up a name for it—model-dependent realism.

Stephen thought that names were important, even in physics. He thought the term "black hole" was brilliant, and that the general public would have been much less fascinated by those objects had they been called, say, "gravitationally completely collapsed objects," which is how people sometimes referred to them before Wheeler came up with "black hole."

To justify my proposed name, I brought a philosophy-of science textbook to the office one morning, planning to discuss the realism versus anti-realism debate with Stephen before making my suggestion. But the lack of interest on his face when I showed him the text convinced me to ditch that idea. So I tossed the book aside and simply blurted it out: model-dependent realism. He liked it, and so that's the term we used. To me it meant accepting different theories for different applications—and accepting, for all practical purposes, different realities. That's how I understood Stephen's belief in the pills. It was a model that hadn't been proved, but neither had it been disproved, and the reality it represented appealed to Stephen.

With Stephen's pill regimen finished, I finally got to pop my question.

"Why do you want to write that philosophy is dead?" I asked him. "It's not dead," I said. "What used to be called 'natural philosophy' is dead, but not philosophy."

Natural philosophy was a precursor of the sciences, a branch of philosophy in which scholars attempted to understand nature through pure reason rather than reason plus experiment. It was rendered archaic by the development of the scientific method. Stephen knew all this, but I continued making my case.

"I agree that today we can understand the universe better through science than philosophy," I said. "But there is also the philosophy of life. There's ethics. There's logic. There's the philosophy of individual disciplines such as mathematics and physics. Those branches of philosophy aren't dead."

Stephen gave me a critical look. He obviously saw things differently. As I awaited his response, I gazed idly at him. Suddenly it struck me that the sport jacket he was wearing was a couple of sizes too big. He seemed lost in it. And in his slacks, too. I supposed that well-fitting clothes would be difficult to find for him, given his almost complete lack of muscle. There wasn't much meat on his bones.

"I have an idea," I said to Stephen. He stopped composing

his answer and turned his eyes toward me. "How about writing 'As a way of understanding the physical world, philosophy is dead'?"

Stephen grimaced. He looked back at his computer screen and resumed working on his response.

Impatient for his reaction, I got up and walked around to watch as he typed. It felt a bit funny, as I wasn't accustomed to doing that—and I'd been warned that he "generally" didn't like it. But on this occasion he didn't seem to mind. In time, he'd grow comfortable with my pulling up a chair to sit beside him. He would come to welcome it because it would speed things up—I could watch his sentences evolve on his screen, and sometimes I could finish them for him or guess where he was going. When I got it right it would save time, because he didn't have to finish typing it all out. But if I guessed wrong, he'd be annoyed. If I guessed wrong twice in a row, he'd be *really* annoyed.

I got to his side as he was just finishing his typing. His text read, "Your sentence has no punch."

I answered before he had a chance to make his computer voice it.

"That's true," I said. "It's less punchy, but to say 'Philosophy is dead' is an oversimplification."

He looked back at his computer screen but didn't type anything new, just had his computer give voice to what he had already written. "Your sentence has no punch," it said.

"I do get what you're saying," I told him. "But if we say philosophy is dead we will piss off a lot of people."

He looked back at his screen and clicked again. Stephen could control the volume of his computer voice, and now his computer voice repeated "Your sentence has no punch," this time very loudly.

I looked at him. He bent his mouth into an unnatural shape,

like an exaggerated smile, but upside down. He was apparently frustrated that I wasn't getting the point. I had to admit that my phrasing really didn't have much punch. And if there is one thing Stephen liked, it was punch.

There were two types of coworkers that Stephen had little patience with: those who were not bright enough to understand his point, and those who didn't accept his point. Was I being too literal? Was I treading on his flair for the dramatic? Don Page, one of Stephen's students from the early days when he could still control his wheelchair, told me that once, during an argument in which Don wasn't conceding, Stephen charged him with the chair and would have run him down if he hadn't jumped out of the way. When I knew him, Stephen no longer had control over the chair. He couldn't run me down. But this issue wasn't worth making him wish he could.

"Okay," I said. "But we'll cause a stir."

At that thought, his frown flipped into a big smile. He liked causing stirs.

Years later, when the book came out, we'd learn that his instincts had been right. The sentence was applauded by many of our readers. But I'd been right, too—it pissed off many others. Especially philosophers.

Not many people go into physics giving much thought to the deeper meaning, if any, of what they are doing. But older physicists often do give it thought. Their experience with physics drives them to gradually develop their philosophy and their attitude toward the meaning of their work. That's how it was with Stephen. His physics didn't grow from his ideas about model-dependent realism. His ideas about model-dependent realism were an outgrowth of his physics.

Stephen's career started after his Ph.D. dissertation, written in 1966, when he was twenty-four. In that work he showed that Einstein's general relativity required that the universe had begun with a big bang. That made him famous in the cosmology world, but not yet the dominant figure he would later become. His dominance grew out of his next project, in which he combined general relativity and quantum theory, conflicting theories that presented vastly different conceptualizations of the universe, of the nature of space and time, of force, of motion, even of the sense in which the present affects the future. It was in his embrace of the contradictions in those two theories that one sees the roots of his ideas about model-dependent realism. And in deftly moving back and forth between those two theories, he became the first to apply both to the same important physical process, thus leading the way for others. That work was his research on black holes, culminating in his discovery of what is now called Hawking radiation.

6

Before I'd started my visits to Cambridge, Stephen and I used the occasion of his annual visits to Caltech in 2005 and 2006 to create the plan for what our book would include. I wasn't used to that level of planning. In my solo career, my editor at Pantheon, Edward Kastenmeier, had given me a lot of leeway. It might cause a problem if I'd said I was writing about quantum computing and then turned in a treatise on women's soccer, but short of that, he was pretty flexible. I'd start with a rough outline and then figure things out as I went along. Stephen wrote his breakthrough book, *A Brief History of Time,* with even less of a plan. In contrast, we had plotted out *Grand Design* as if it were a physics research paper.

Stephen had apparently wanted to nail everything down before we started to write. But nothing was ever final. We'd make a decision on something one day and then revisit it a few days later—as we were still doing. I was beginning to think that we'd debate the contents of the book forever and never write anything. Then one day, when we were having a long, slow afternoon beating a few dead horses, out of the blue Stephen said, "It's time to stop talking." At first I wasn't sure what he meant. Was it dinnertime? But he wasn't talking about red wine and pot roast. What he'd meant was, it was time to start writing. There were still things to fill in and still things we were debating, but he'd apparently had enough. And so began my trips to Cambridge and our cycle of writing, exchanging drafts, and meeting to scrutinize every idea and every word.

In working with Stephen, you couldn't be shy about defending your point of view. On the other hand, you could heatedly contest a point he was making and a half hour later be laughing with him in a pub. A few years before Stephen's death, Neil Turok, a leading cosmologist and friend of Stephen's, would write a series of papers criticizing some of the work Stephen was most proud of. It didn't affect their friendship. That's the culture of theoretical physics. When someone identifies a flaw in your reasoning, even if they say *I believe you may have been laboring under a slight misconception that possibly impacts the validity of your otherwise brilliant argument,* it can feel like they've said *You're an idiot.* But deep down, you know they are doing you a favor—if an idea is not going to work, it's better to see it before you and possibly others waste any more time marching down a dead end street. As a theorist you know that most of your ideas will be proved wrong—after all, if the majority of our ideas worked, all the open problems of physics would already have been solved. So there's no beating around the bush about wrong ideas—and no hard feelings.

As we were writing, Stephen's experience with *Brief History* was always in the background, sending us cautionary messages. With its discussion of concepts such as light cones and imaginary time, *Brief History,* despite all that editing—and despite its popularity—could be tough going. Once, when we were eating at Burger Continental, a cheap hamburger joint near Caltech, a student approached and told Stephen *Brief History* was one of his favorite books. Stephen replied, *Thanks, but did you finish it?*

He believed that most people never got to the end, so he always made the same reply. It was one of those canned sentences he had stored on his computer because he had so much occasion to use it. He wanted *Grand Design* to be different. We knew there would be parts of the book describing advanced physics that most people would have trouble with, but we wanted people to finish it.

I was now in Cambridge on another critique-each-other trip. We'd just finished dinner at Stephen's house. As usual, his wife, Elaine, wasn't around and white-haired Joan had done the cooking. The look on Joan's face made me think that the standing she had to do while cooking was causing some back pain, but she was always cheerful and never complained.

This time she'd made lamb stew, with lots of gravy. Gerald, who was Stephen's carer that night, mixed in the liquid when he chopped up the meat for Stephen. The gravy made it easier for Stephen to swallow. As always, the serving plates were piled high. In addition to the lamb, there was mint jelly, greens, and mashed potatoes, one of his favorites. Afterward, berries with clotted cream. And wine, a bottle that I picked from his wine closet as usual, this time based on its pretty label. It wasn't bad. Not that I'd know.

Stephen took his wine from a glass, but a spoonful at a time. It didn't usually add up to much. But this night he went a bit overboard. Me too. Maybe it was the lamb. We ate, as always, at the table in the dining area, which opened on one side to a living room, with a patio just outside. On the opposite end was a galley kitchen, where Joan was starting on the dishes. Semiretired now, Joan still helped Stephen out in a variety of ways. She was devoted to him, and Stephen and his carers all loved her, too.

In some ways, it struck me, this was also Stephen's family. He saw his daughter, Lucy, on Sundays, and they were close. He seemed to rarely see his younger son, Tim. His older son, Robert, lived in Seattle. At this point in their relationship, Elaine would come and go like a hummingbird just passing through. But Joan, Judith, and his carers saw Stephen and looked after him around the clock. They dined with him, put him to bed, took him to doctors, traveled with him around the world, and tended to his needs on all those rainy days.

When his carers had a family event, a birthday or anniversary, Stephen would often attend, and when they needed things, he'd sometimes give them the money to buy them. He lent one carer the money to buy a car. He promised the daughter of another that when she got married, he'd provide the fireworks. Back in his teens, Stephen and his friends used to make their own fireworks, and he still loved them. He often had them at his own big parties. Not amateur stuff like he used to make, but the kind they set off in stadiums. Sometimes someone would call the police, and they'd come and tell him to stop. Then the police would leave and Stephen would have the fireworks start up again.

Tonight's dinner had been a feast, but I felt bad. Joan looked exhausted. After a while she left, and Gerald retired to the next room to read. Along the way Gerald turned on the television news. Stephen liked to watch the news, which I found odd because it often annoyed him. Right now he was making a face at a report about something or other that Parliament had voted on. I asked him if I should turn it off. He raised his brow, meaning *yes,* so I did.

We sat there looking at each other in silence for a minute. I suddenly understood why he liked my company in the evenings, how, after he was fed, his nights could quickly turn lonely. His eyelids drooped. I wondered if the wine was about to put him to sleep. And then he seemed to perk up, as if a thought had suddenly popped into his mind. He started to type.

"Are you healthy now?" he asked.

I nodded. After my last visit I'd undergone surgery for a small-bowel obstruction, and then one day, a while afterward, I'd fainted. At the hospital they said my blood pressure was 58/30. I'd had massive intestinal bleeding. They stuck an intravenous line in me and started infusing what would eventually total thirteen units of blood. That's a complete oil change. But the bleeding came and went in spurts, and even after a dozen tests and procedures they couldn't pinpoint the vessel that was causing it. One night while in intensive care I overheard my doctor telling a resident to keep an eye on me because I was in danger of "bleeding out" before morning. He'd apparently missed that medical school class on keeping your voice down. He also misread his crystal ball. After ten days my bleeding fits stopped as suddenly as they'd started.

Lying in that hospital bed, having heard the prediction that I might soon die, I'd thought about many things. About how I might have seen my family for the last time. How I'd never know whom my kids married or how their lives would turn out. How I'd no longer be there when they needed me. They were still so young—would they forget me? Had my life had any meaning?

My mind summoned random images, of ocean waves, a sunny beach, mountains covered with a blanket of snow. They were clichés but I didn't want to let go of them. I looked out the window and thought about the great beauty of the blue California sky and the palm trees outside. Had I taken it all for granted? Should I have stopped more often to appreciate them? Was it now too late?

I wondered whether Stephen had such thoughts. In his many moments at the edge of death, did he think of the sky, or the stars he loved to gaze at? Did he want to hang on for the sake of his children? Did he have regrets? In time I realized that he'd had so many life-threatening incidents—a stoma breakdown, a lung infection, a sodium imbalance—that being at death's door became routine. The plagues kept coming, but he'd made peace with his life and oft-expected death. He'd worked through all those thoughts that I, plunged into my first life-and-death crisis, suddenly had to confront.

When I was in that intensive care ward and feeling more vulnerable than I ever had, I thought of how vulnerable Stephen had seemed to me, and how wrong I'd been about that. It occurred to me that Stephen had proved himself to be an iron man in a fragile man's façade. Science is about observation and evidence, and regardless of appearances or the pronouncements of doctors, the evidence suggested that nothing could fell Stephen. I, on the other hand, had proved to be fragile. Lying in my hospital bed, I'd thought about how ironic it would be if, of the two of us, I was the one who died before finishing our book. I had a dream. It was fuzzy, but in it Stephen and I were in a two-person race. I took off, happily sprinting, while he followed, pushed along slowly by his straining wheelchair motor. And then I toppled and lay still on the track as Stephen rolled past with his eyebrows raised and a smile on his face.

It seemed odd to me to have such thoughts and dreams on my potential deathbed, but I'd had them. I told Stephen that, and he appeared amused.

"You like to make bets," I said. "We should bet on which of us will go first."

He frowned.

"Why not?" I said.

He started typing.

"The loser won't be around to pay," he said.

He had a point. I sipped some wine and asked if he'd like me to wipe some drool that had collected on his chin. Another yes.

Over time the carers grew comfortable letting their guard down a bit when I was there, and I'd gotten used to doing minor carer-like tasks for him. I noticed that Stephen liked the attention. Sometimes, I thought, he asked for an adjustment to his sitting position or his glasses just to have human contact. He seemed to like to be touched, and I could understand his being hungry for it. He slept alone and couldn't snuggle or trade caresses with his lover. He couldn't even hug friends when greeting them or reach out to shake someone's hand.

Stephen had gotten word of my bleeding incident and sent a kind get-well card signed by him, Judith, and several others. I thanked him for it. "Funny how being in the hospital made me think a lot about death," I said. "Never thought much about it before."

His look seemed to say welcome to the club.

"I know. *You* go through life-and-death scares all the time."

He raised his brow. Yes. He started typing. "Then back to physics," he said.

Physics. With Stephen no conversation was ever far from physics. "Doesn't it frustrate you that you can't write the equations?" I asked.

He grimaced. I wasn't sure whether he meant no, it didn't frustrate him, or that he didn't like my prying. The issue was something I'd wondered about but hadn't felt close enough to him to bring up. Now that I had, I hoped I hadn't crossed a line.

He typed. "My disability was gradual," he said. "I had time to adjust."

"Imagine the physics you might have done if you hadn't had your disability," I said.

His frown said he disagreed. He started typing. It took a while, but I didn't peek. "It helped. It helped me focus," he finally said.

I marveled that he could see anything positive in his plight. And I marveled at his passion for physics. He'd met my son Nicolai on a couple of his trips to Pasadena. I told him that

Nicolai, a basketball player who worked at it for hours each day, would always say, "Basketball is life." I told Stephen it was amazing that, after so many decades of life and the ordeals he'd gone through, he still had that same passion. "For you, physics is life," I said.

He wrinkled his nose. He disagreed again. He started to type.

"Love is life," he said.

Stephen's comment that love is life touched me, reminding me that though his disability was a barrier to developing both emotional and physical relationships, he thrived on human connection as much as anybody. Still, his comment surprised me, because in his choices he'd often chosen physics over human connection. Even before he became wheelchair-bound, he'd make himself essentially incommunicado for days on end while he puzzled out some important issue in his research. He hadn't spent a great deal of time with his children when they were growing up. His wife, Jane, had felt neglected. Despite all that, I had no doubt that his family and relationships were central to his happiness. It struck me again that in his life, as in his physics, he embraced many contradictions.

To Stephen, contradiction in physics meant opportunity, that there was something that begged to be reworked, reconciled, or simply understood and accepted. As he moved on from his Ph.D. work, his capacity for physical adventure was fading but his adventures in physics were becoming ever braver and more audacious. In the years to come he would move on from his work on the origin of the universe and join that band of adventurers who were starting to explore the strange world of black holes.

Even among those black hole pioneers, Kip Thorne told me, "Stephen was an unusually bold thinker." That was saying something. It was like being an unusually heavy sumo wrestler, or an unusually wet fish. Because in the early day of black hole research, when our knowledge of those strange and exotic objects was relatively new, no one in the field was timid. They couldn't be. They were trailblazers in a world ruled by some of the strangest implications of relativity—that there is no universal "flow" of time, and no "present time" that is shared throughout the universe. We seem to experience that things "exist" in a common present and that events occur one after another, but, as Einstein said, "For us believing physicists the distinction between past, present, and future only has the meaning of an illusion, though a persistent one."

The theory of black holes even allows for time travel. It predicts that if you fly to a black hole, hang out for a while, and return, when you get back you might find that you've jumped a few hundred or thousand years into your home base's future. Do it repeatedly and you could watch civilizations rise and fall, as if you were watching the future of your planet on fast-forward. Today the science-fictionish world of black holes is more familiar, but back then, before every schoolkid knew that space could be curved, it was all new and mind-bending. And even among that band of daredevil thinkers, Stephen stood out.

Stephen's first contribution to black hole physics concerned the black hole horizon, a key concept in the definition of those exotic objects. In colloquial terms, physicists have always thought of a black hole as a region in space that, due to its immense gravity, allows nothing to escape to the outside. One can think of that region as being defined by its horizon.

In the words of Roger Penrose, the black hole horizon is "the outermost location where photons [light] trying to escape the hole get pulled inward by gravity." The name comes from an analogy—just as we on our planet can't see the sun after it passes our earthly horizon, an outside observer cannot see past the horizon of a black hole.

In his work on black holes, Roger Penrose turned his definition into a precise mathematical form. His formulation sounded reasonable and soon became the standard. But as Stephen studied the physics of black holes, he realized that the Penrose horizon was actually, as Kip put it, "an intellectual blind alley."

Two issues tainted the Penrose approach. The first is an issue that touches the soul of relativity, the contradictions that arise among disparate observers. According to relativity, different observers—depending upon the strength of gravity in their environment and their movement in relation to each other—may not agree on the size and shape of regions of space and durations of time. This can make one's analysis a mess. But there is a way around that: researchers can stick exclusively to concepts that are defined in a manner that is independent of the observer. To do so brings multiple blessings. First, it ensures that the laws and phenomena they discover apply to everyone. Also, it makes the mathematics simpler. Finally, and perhaps most important, it greatly enhances our ability to interpret the equations. According to Penrose's definition, however, a black hole's boundary would *not* be the same for all observers. Someone falling into the hole, for example, might see the Penrose horizon differently than would someone hovering outside it. So which is it? You'd have different horizons, depending upon who was looking.

The other problem in Penrose's approach is that, if defined his way, the horizon can make discontinuous jumps. For example, when a glob of new matter falls in and the black hole becomes larger, the horizon will suddenly grow in size. In complex situations, such as when two black holes collide, those jumps can be bizarre and difficult to work with.

Penrose was aware of those drawbacks but stuck with his definition. Sometime in the early months of 1971, however, Stephen came to realize that a more productive way to think about the horizon of a black hole is to view it as a region in space-time, rather than as a region of space *at a given time,* as Penrose had done. So Stephen redefined the horizon as the boundary, *in both space and time,* beyond which signals such as light rays could not be sent out to the distant universe. He proved mathematically that this definition remedies both weaknesses of the Penrose approach—the black hole boundary would be the same for all observers, and it would always change smoothly and never jump around.

How can one understand the difference between the two definitions? Imagine a black hole with an immense mass of debris around it, outside it but imploding and soon to be swallowed up.* Imagine also a tiny rocket ship that is just outside the black hole, trying to escape its pull and leave the area. The rocket is firing its jets to propel itself away. Since it is outside the black hole, one might think it could be successful at that. However, the moment the immense mass has been swallowed up the black hole will grow larger, and if the mass is great enough, the black hole will then encompass the rocket, so that the rocketeer cannot escape.

In Penrose's description of those events, the rocket is at first outside the black hole horizon. A moment later, as the immense mass falls in, the Penrose horizon jumps outward and the rocket is trapped inside. So though it was not possible for the rocket to

* For technical reasons, that mass should take the form of a spherical shell, but that detail is not important for our purposes.

escape the black hole, the Penrose horizon did not reflect that at first; it did so only after the immense mass had fallen in.

The way Stephen defines the horizon, if the rocket is fated to eventually be swallowed up by the enlarged black hole, this will come as no surprise, because the Hawking horizon will encompass the rocket from the start. The Hawking horizon, in other words, grows *before* the immense mass falls in. It depends not only on the present state of things, but on what will happen in the future. It thus violates the laws of cause and effect: in this case, the effect (the black hole's horizon growing) precedes the cause (the mass falling in).

Stephen's definition of the horizon requires that you know the full history of space-time, including its entire future, though in practical applications objects far away in space and time can be ignored. Physicists call a definition in which something is determined by future events teleological. They borrowed the term from philosophers, who use it to refer to the explanation of a phenomenon in terms of its eventual purpose rather than an immediate cause.

Philosophers have pondered teleological laws of nature at least as far back as Aristotle. Science teaches us that rain falls because the moisture in clouds condenses into droplets of water, which are heavier than air. But in Aristotle's view there was another reason: it rains so that plants can grow, in order that people can eat them. The demands of the future, he believed, shape the present. We all make life decisions that way, to a greater or lesser extent, depending upon our personalities. For example, if you're offered a slice of cheesecake at lunch, instead of reacting according to your current desires, you might take into account what you're going to be eating for dinner. In physics, however, forces act and objects react according to present conditions, so although teleological concepts are part of our everyday lives, they are rarely used in physics. Stephen's teleological definition of the horizon was a gem of creativity. It was his boldness that enabled him to embrace and explore the idea, unlike others, such as Penrose, who had quickly discarded it.

It may seem that the nuances of how theorists define the black hole horizon shouldn't matter much because the definition of a term is a choice made by physicists, not a statement about nature. But the concepts we invent influence the ideas we generate and the conclusions we draw. In time, Stephen's definition proved to be a powerful advance and was widely adopted by others. It guided their intuition and shaped their mental pictures of the processes black holes undergo. Stephen called his version of the horizon concept the absolute horizon and Penrose's version the apparent horizon to distinguish between them. With that redefinition, Stephen had redefined not just the horizon but the way theorists think about black holes.

Armed with his new way of thinking about black holes, Stephen worked furiously to understand what general relativity tells us about the laws that govern them. He'd seal himself off for days at a time. Jane would approach to talk about some quotidian issue that had come up, but he'd stay in his physics world. She'd approach for reassurance that she was important to him, and he'd ignore her. He would blast Wagnerian opera for hours at a time on the record player as he worked, just as he'd done after receiving his diagnosis, just as his parents used to do when he was a child. Jane grew to hate Wagner. She thought of him as an "evil genius," an alienating force in their marriage.

Whatever Wagner subtracted from their relationship, he apparently added to Stephen's physics. After working for a year and a half and collaborating with two colleagues, he achieved a second important breakthrough, the laws of black hole mechanics. Formulated in August 1972—when Stephen was just thirty—this was a set of rules that govern how black holes grow when matter falls into them and what happens when black holes interact with other black holes.

Stephen's laws were ahead of their time. The first experiments demonstrating indirectly but with high confidence that black holes even existed wouldn't come until around 1990, based on observations of a heavenly body called Cygnus X-1. The first direct observations of the disturbances of space-time that are the signature of black hole collisions—a certain type of gravity wave—didn't come until 2015, in the LIGO* experiment for which Kip Thorne shared the Nobel Prize. The first (almost) direct observation of a black hole didn't come until 2019, the year after Stephen died.

Despite our inability at the time to see a black hole, Stephen believed that black holes could provide unique insight into the nature of gravity, space, and time, revealing secrets that don't make themselves apparent in ordinary circumstances. His intuition would prove correct.

The discovery of the laws of black hole mechanics was an important step toward understanding those esoteric objects. But the laws had a strange peculiarity that would also turn out to be important: They looked a lot like the laws of another field, thermodynamics, the physics of heat. Each black hole law, in fact, was identical to a thermodynamics law if you replaced certain terms from black hole physics with a corresponding term in thermodynamics.

Consider one of the black hole laws in particular, the area

---

* LIGO is the Laser Interferometer Gravitational-Wave Observatory. The announcement of the discovery came in 2016.

increase theorem. It says that in any interaction black holes undergo—whether they are coalescing, swallowing matter, colliding with each other, and so on—the sum total of the areas of all the black hole horizons will always increase. That's not true of ordinary objects. For example, suppose you take two identical balls of clay, merge them, and from that glob form a new ball. Simple high school math will then tell you that the surface area of the new ball will be about 20 percent *less* than the sum of the original surface areas. But due to the curvature of space, if you merge two black holes, the horizon—the analog of the surface area—will be *greater* than the sum of the originals.

Physicists immediately noticed that the area increase theorem is strikingly similar to what is known as the second law of thermodynamics. The area increase theorem says that in any black hole interactions, the sum of the black hole "horizon areas" always increases. The second law of thermodynamics says that, no matter what physical interaction takes place, the entropy (degree of randomness) of any closed system always increases. Just replace the term "horizon areas" with the word "entropy" and the black hole law becomes the thermodynamic law.

Virtually all physicists believed the correspondence of these laws was a weird but meaningless coincidence. But a graduate student at Princeton named Jacob Bekenstein didn't agree. Bekenstein speculated that the connection between the laws should be taken literally—that the entropy of a black hole is proportional to the surface area of its horizon.

Entropy is a measure of disorder. In an ice cube, for example, the water molecules are arranged in orderly hexagonal rings, while in liquid water the molecules bounce around at random. An ice cube, therefore, has relatively low entropy, and its entropy increases when it melts. More generally, a low entropy system is an orderly system, or a simple system in which there aren't many constituent parts that can become disordered. A typical high entropy system, on the other hand, is a complex system in disarray.

A black hole seemed to be too simple to ever be in a state of disorder. Once formed and in a settled state, a black hole in empty space was thought to be like a billiard ball. With no constituent parts, there was nothing that could ever be in a state of disorder. As a result, black holes were said to have no disorder at all—and hence an entropy of zero. Bekenstein's ideas contradicted that picture and were met with widespread derision.

There was another reason people found Bekenstein's theory unpalatable. According to the laws of thermodynamics, anything with an entropy greater than zero must have a temperature greater than zero—it cannot be completely cold. And anything with a temperature above absolute zero must emit radiation—it will glow.*

That's a problem, because when something glows, it releases—radiates—energy. That energy would have to come from the black hole's mass. A glowing black hole, in other words, would be slowly converting its mass to electromagnetic energy (according to Einstein's famous $E = mc^2$) and radiating it away.†
As a result, the black hole would gradually shrink and eventually disappear completely. It would, in a sense, "evaporate," as everything inside eventually leaked out in the form of radiation.

Today we know of that radiation as Hawking radiation, but back then, ironically, Stephen didn't believe in it, or in any of Bekenstein's ideas. They contradicted the picture of black holes that Stephen and others had painstakingly derived from the equations of general relativity. Bekenstein recognized that his

---

* Though not necessarily in the range of visible frequencies.
† The story is a bit more complicated for black holes with spin, but that is outside the scope of this discussion.

theory of black hole entropy seemed to demand that black holes radiate. Yet he also shared the opinion that black holes cannot radiate. He didn't know how to resolve that issue, but he stuck with his idea that black holes have entropy.

The widespread attacks on Bekenstein illustrate why it takes courage to advocate for new ideas in physics. If you have convincing evidence, chances are you'll win your battles. But though Bekenstein believed that black holes have entropy, he didn't go far enough and accept the consequence of black hole radiation. Nor could he adequately defend his ideas. Virtually no one accepted his point of view. He got shot down, and Stephen was one of the lead attackers.

As it would turn out, when the old theory of black holes, based solely upon general relativity, was amended to incorporate the principles of quantum theory, the idea that black holes have entropy would be confirmed—and it would be Stephen himself who, admitting that he'd been wrong, would reluctantly flip sides in the argument and prove Bekenstein correct.

With all that wine and lamb and the talk about my near-death experience in the hospital, by the time I left Stephen's home it was after ten. Still, I wasn't ready to go to my room at ancient Caius college. Given Stephen's work schedule, I tended to stay out late and return to the room only to collapse in bed.

This was winter, and my room, with its stone walls, small windows, and low ceiling, was tiny and dark. I suppose it had some charm, especially if you're a bat. But I wasn't in the mood to stare at the ceiling, so I walked the half hour or so from Stephen's house to a pub I knew. The pubs in Cambridge were supposed to close at eleven, but "close" meant different things to different people. To the proprietor of this pub, a fortyish Chinese woman, it meant close your doors. That she did—she locked them tight promptly at eleven. What she and her British bartender husband didn't do was ask their customers to leave. Instead, they kept serving until everyone had slowly filtered out. Sometimes that wasn't until two, or even later. It was a business model of dubious legality, but it worked just fine.

Cambridge pubs were unlike any I'd encountered. The half-inebriated person guzzling pints next to you wasn't just a drunk. He or she might also be an astrophysics graduate student or a renowned neuroscientist. One of my favorite evenings was spent listening to a beer-filled fellow talk about the agricultural economics of West Africa. Not my usual choice of topic, but it went surprisingly well with a few stouts and mixed nuts.

On this evening I made the mistake of letting the bartender steer the subject to my work with Stephen. Since I was a regular there, he and his wife knew I worked with him, but I usually managed to avoid the subject. Not this time. This time he poured me a free beer, then told me the price was that I'd have to tell him about black holes. I hated getting into such situations. I was there, after all, to forget about black holes. Now I'd been sucked back in.

Luckily, the bartender was one of those who'd rather talk than listen. He asked me a question, and just as I started to answer, he picked it up from there. And for the next twenty minutes he proceeded to tell me everything *he* knew about black holes. Most of it was correct, too.

We'd come far, I thought, from Stephen's early days, when black holes were an exotic subject. Back then, few physicists cared to discuss them. Now you can get a lesson from your bartender. As he droned on and his wife occasionally rolled her eyes, my thoughts drifted to how Stephen was the person most responsible for that, how he'd had a huge effect, not just on the culture of physics but on the culture at large. He appreciated that, perhaps especially now, in his later years. For the questions he'd wanted to answer were questions not only for physicists, but for all of us. I realized then that if his discoveries in physics lent him a kind of immortality, so did the physics he shared with the public. That thought added to a feeling I'd had since my surgery that Stephen was indestructible.

7

The 1970s were not a good decade for Stephen Hawking's body. He'd matured as a physicist but his disability was also advancing. In the early 1970s he lost most control of his hands. He'd drawn his last diagram, written his last equation. In 1970 he could propel himself using a four-legged walker. By 1972 he required a motorized wheelchair. By 1975 only those who spent considerable time with him could understand his garbled speech. He was then thirty-three years old.

Stephen didn't know then that he'd be a survivor. He expected the 1970s to be the decade of his death. He could think and feel but hardly move. His wheelchair became his throne, and although his faith in himself was strong, he needed no proof that his creeping decay foreshadowed his fate. He knew he could last only until his disease pummeled the muscles needed for breathing. Then he'd begin suffering bouts of pneumonia, and in the end he'd surrender in a fit of suffocation. All this he expected to happen in the next handful of years. Stephen had dreams about what he wanted to accomplish in physics, but about his fate he was no dreamer.

Aware of his finiteness, Stephen resolved to answer the

questions that inspired him, to understand, before he ceased to exist, the meaning of existence. He realized, though, that he could not continue to work as other physicists did. He had to alter his approach, change his style. He refused to succumb to his disability, but he did adjust to it. In his personal life he began to develop the sophisticated nonverbal communication that, among those who knew him well, he'd become famous for. In his professional life he altered his approach to the mathematics of his theories in two distinct ways.

The first had to do with the mathematical approximations he considered acceptable. Galileo eloquently argued that the book of nature is written in equations, but he neglected to mention that they are equations we can't solve. Newton's theory of gravity famously explains the orbits of the planets, but we can solve his equations only for an unrealistically simple solar system with just one planet. In the much-celebrated quantum theory of atoms, all of chemistry emerges from a single equation, but the only element whose behavior we can calculate *exactly* via that equation is hydrogen—the simplest of them all. If we want to describe the planetary orbits in our actual solar system or derive the chemistry of elements other than hydrogen, we have to settle for an approximate picture, the mathematical equivalent of educated guesses. Those approximations and guesses do not come with any mathematical guarantee of correctness, but working physicists develop a good feeling for what is legitimate and what isn't.

In physics we accept mathematical manipulations that we think "ought to work," while mathematicians have the inconvenient habit of requiring proof. As a result, they sometimes complain that physicists abuse their subject. And it's true—in our attempts to uncover the truths our equations conceal, we break mathematical laws, evade the math police, and ignore their court orders. We chop off bits of our equations to subdue them, then interrogate them and assume that what they confess is close enough to the truth that we can piece it together. In all but the simplest investigations in theoretical physics we alter, assume, and approximate, and then we argue why our simplified model and the conclusions we've drawn from it are nevertheless valid. Sometimes they are, sometimes they aren't. The argument over that is part of the professional physicists' (sometimes heated) dialog, which is a lot messier than the stereotypical portrait of science. Yet the fact that our airplanes fly, our lasers shine, and our computers compute suggests that a lot of our muddling works out in the end.

Different theorists have different degrees of tolerance for argumentation that has holes or loose ends or that otherwise involves questionable mathematical manipulation. Some have a taste for more rigor, others for less. The former publish only when they find strong justification for their arguments; the latter play it looser. Early in his career, Stephen was one of those with a strong taste for rigor. But then he changed. Believing that the end was near—his end, that is—he made a concession, and starting in the early 1970s he began to take more liberties. Crossing t's and dotting i's takes time, and he didn't have time to spare. "I want to make as much progress as I can, and I can't if I'm rigorous," Stephen told Kip. "I'd rather be right than rigorous."

The other adaptation Stephen made was to think more geometrically, in pictures rather than in equations. Much of physics can be viewed as geometry. It doesn't have to be viewed that way, but it can be. The relation of the more and less geometric approaches is something like that between the geometry and algebra you study in high school. In geometry class, you're presented with lines, angles, circles, triangles, and other shapes, and you're given rules and asked to reason with them. In algebra class you work with many of those same concepts, but in the form of equations—for example, the equation for a line, a circle, and the sine and cosine functions. You can prove your theorems one way or the other. That can be true in physics as well. Especially in relativity, which, as Minkowski had demonstrated, is particularly suited to the pictorial, geometric perspective.

Stephen compensated for his inability to write equations by developing a sophisticated geometric language that allowed him to envision the physics he wanted to study through pictures in his head. He gradually trained himself to mentally manipulate curves and intuitive diagrams rather than equations on the whiteboard. He had always thought differently from other physicists, but he had now also developed a unique language, one that was his alone.

For some problems, Stephen's language was more powerful than traditional equation manipulation. In these cases, rather than his disability acting as a handicap, it had forced him to develop a superpower. He could see what others didn't and gain insights they couldn't. For other problems, his approach was less powerful than that taken by others. He learned to recognize which problems were which and focus on those issues in which he had an advantage. For those questions Stephen had, in Kip's words, "a power that nobody else could begin to match."

The road to Hawking radiation went through Moscow. Stephen landed there in September 1973, when he and Jane accompanied Kip there with the goal of meeting some brilliant Russian physicists whose freedom to travel had been restricted by the Soviet government, either because they were dissidents or because they were Jewish. The physicists couldn't visit Stephen in Cambridge, but they did make the pilgrimage to Stephen's two-room suite at the Rossiya Hotel, off Red Square. On one of these visits, Stephen learned of a peculiar conjecture made by his host Yakov Zel'dovich.

When a person dies and is cremated, whether that person was fat or thin, tall or short, beautiful or unattractive, kind or evil, ignorant or educated, his or her body is reduced to a pile of carbon ash. People are individuals, but carbon is carbon. All that might distinguish the remains of a fat king from that of a slim ballet dancer is the mass of that pile. Stars greater than a certain size meet an analogous fate.*

When a massive star dies, when it implodes to form a black hole, virtually all traces of its former identity are lost. The elements and particles it was made of, the state of the turbulent plasma within it, the layers of structure it had developed, all vanish from existence. After its collapse, all that remains of the star's past individuality are the values of the three parameters that are the only traits a black hole may possess: mass, spin, and electric charge.

Much of the popular lore about black holes, and a lot of the work in black hole physics, focuses on the simplest black holes, those with zero charge and spin. The only characteristic those black holes possess is mass. But Zel'dovich's speculation concerned spinning black holes. He proposed what was then an odd idea—that spinning black holes should radiate energy.

According to Zel'dovich, the radiated energy would be drawn from the black hole's spin. Over time, the radiation would drain that spin energy and the black hole would spin more slowly, until it eventually stopped both spinning and radiating.

Unlike Bekenstein's theory, the idea that a spinning black hole would radiate energy was not revolutionary, because the

---

* Only very massive stars have enough internal gravitational pull to form black holes upon their collapse. Our own sun, being relatively light, will die a quieter death, ending its life as a white dwarf.

energy radiated would be drawn from the black hole's spin and not from its mass. A spinning black hole could maintain its mass and still radiate. It wouldn't have to shrink and eventually disappear in the process.

Zel'dovich published his ideas, but his argument was complicated and involved some questionable mathematics. His paper had been ignored and largely forgotten. In that room off Red Square, however, Zel'dovich explained his theory to Stephen, who was intrigued. Zel'dovich's analysis depended upon the laws of both gravity and quantum theory. Ideally, he would have used a theory of quantum gravity to conduct the work, but since there was no such thing, he attempted to carefully employ elements from both general relativity (for the gravity aspect) and elementary particle physics (for the quantum aspect). Stephen was skeptical of the way Zel'dovich had done it, so he decided to investigate the process using his own geometric methods.

When Stephen did the analysis his own way, he found that Zel'dovich was indeed mistaken—but not in the way Stephen had suspected. Stephen's work confirmed that spinning black holes radiate energy, but it showed that black holes that didn't spin also radiated energy. According to Stephen's calculations, *all* black holes radiated, just as the Bekenstein black hole entropy theory seemed to demand.

At first Stephen thought he'd made a mistake. Maybe one of the approximations he'd used to do his calculations wasn't valid. But he couldn't find a problem in any of them. And when he calculated the characteristics of the energy radiated, he found that they were what you'd expect if Bekenstein had been right.

Bekenstein had insisted that black holes had nonzero entropy. That had rubbed everyone the wrong way because according to thermodynamics, entropy meant radiation, and people believed that black holes did not radiate. That belief was based on general relativity, an analysis that ignored the effects of quantum theory. What Stephen discovered was that you couldn't ignore it, that quantum theory changed things in an essential way. He showed that when you take quantum effects into account, you're led to expect precisely the kind of radiation that Bekenstein's nonzero entropy demanded and that general relativity could not provide. Quantum theory made it possible for Bekenstein's entropy theory to be correct.

Stephen said that he was annoyed at what he'd discovered and kept it quiet for a while. As he wrote in *A Brief History of Time,* "I was afraid that if Bekenstein found out about it, he would use it as a further argument to support his ideas." But as Richard Feynman used to say, physicists don't tell nature how things behave, nature shows physicists. So Stephen eventually accepted that Bekenstein was correct: black holes have nonzero entropy that is proportional to the surface area of their horizon; they have a nonzero temperature; and they slowly convert the matter and energy they have swallowed up into radiation that they emit back into space, gradually shrinking in the process until they eventually disappear.

Stephen knew that when he announced his discovery he'd encounter the same resistance Bekenstein had and would have to defend his change of sides. Unlike Bekenstein, Stephen had based his belief on the idea on a convincing calculation. But since there was no unified quantum theory of gravity, everyone had their own way of carefully mixing general relativity and quantum theory, and their own way of doing the math. Few were familiar with his geometric approach, so there would be plenty of room to question the particular hodgepodge of the two theories he had chosen to employ. Yet he wasn't afraid of the coming fight.

Stephen decided to unveil his theory of black hole radiation at a conference at the Rutherford Laboratory, south of Oxford. His speech was difficult for those who didn't know him to understand, so his doctoral student Bernard Carr came along. It was February 1974, a cold and gloomy month in England. Stephen wasn't sure a public unveiling was the best idea, but his former Ph.D. advisor, Dennis Sciama, was the conference organizer, and when Stephen told him about his discovery, Sciama was enthusiastic. So were Martin Rees and Roger Penrose. They all believed his result, but then they were all friends of his.

Since his speech was slurred, Stephen's game plan was to read his talk from a transcript that Carr would project onto a screen. That way everyone could follow along. Stephen didn't expect everyone to immediately understand all his math, but he had confidence that his argument was true and convincing. And he knew that he could answer any challenges that might come up in the usual Q&A session after his lecture.

The day was filled with talks. As Stephen was in the lecture hall attending one of them, Jane sat in the tearoom awaiting Stephen's, which was set to start at eleven. A few cleaning women were there on break, having coffee and cigarettes. Though Stephen hadn't yet spoken, Jane now experienced the first reviews of her husband—not of his work, but of his person—from these women.

"There's one of them, that young chap, he's living on borrowed time, isn't he?" one said.

"Looks as if he's falling apart at the seams," said another. She laughed.

It was clear to Jane that they were talking about Stephen. She tried to ignore them, but they made her think. She'd become accustomed to Stephen's condition. However frail he might appear to a detached observer, to her he always looked normal. Sure, whenever he declined she'd notice, but she'd soon assimilate the change, and that would become the new normal. Her ability to adapt was a blessing, for it let her be with her husband without constant thoughts of his coming demise. It allowed her to have hope and dreams of the future. The reminder that, to objective outsiders, Stephen appeared decrepit and near death was like plunging her into an ice bath.

After the women had left, Stephen rolled into the tearoom, oblivious to what had transpired. Like them, he passed up the tea and went for coffee. A few minutes later he was on stage, under the lights, reading his talk from the slides.

I'd heard a story about a seminar at Berkeley. The speaker stood at the front, turning occasionally to scrawl an equation or two on the board as he spoke. There were about fifteen rows of seats, a couple of dozen seats per row with an aisle down the middle. Halfway through the talk a famous professor, sitting front and center, pulled out a pen and wrote the words "THIS IS BS" in large letters on one side of his Styrofoam cup. He held it high over his head and rotated it slightly, back and forth, so that everyone behind him, but not the speaker, could see his message. Then he stood up and without a word walked out.

Physicists can be a tough crowd. Especially when you're arguing against the gospel, and in a language they're not familiar with. Stephen knew that. He himself, in his later years, would engage in analogous antics. Once, when dissatisfied with the talk a postdoctoral fellow was presenting, he disrupted the lecture by spinning his wheelchair in circles at high speed. In those later years, though physically frail, he could stand up to any intellectual tough guy. But those years were still to come. Stephen wasn't yet that fierce, or that high on the totem pole.

In the early stages of a physics career you're more likely to fear than be feared. Another young physicist might have played it safe and declined to give the live unveiling. He or she might have simply sent their research paper off to *Nature,* where it would speak for itself. But if Stephen's voice was weak and shaky, it was due to his disease, not to being cowed by an adverse audience reaction. He might have approached the event with trepidation, but he wasn't going to dodge it.

He read his slides slowly and steadily. Then it was over. There was no applause when he finished. There were no excited murmurs. There were no sounds at all. Had they not followed his talk? Had they lost interest? Or were they disbelieving and appalled? Which was it—had his talk hit them like a tranquilizer dart or a stun gun? As it turned out, it was the latter. The cleaning women's reaction to him would prove to be the kindest criticism he'd receive that day.

The silence ended after a few moments, when the session chairman, John G. Taylor, jumped up. "Well, this is quite preposterous," he said. "I have never heard anything like it. I have no alternative but to bring this session to an immediate close!"

With that, Taylor shut Stephen down. He didn't thank the speaker, as was usual—and there wouldn't be a Q&A, as was also usual.

Shortly after the talk Stephen sent *Nature* a paper—"Black Hole Explosions?"—describing his work. Meanwhile Taylor also sent *Nature* a paper, rebutting Stephen's ideas. Taylor's was accepted by the journal, while Stephen's was rejected. Stephen discovered that Taylor had been the one who rejected his paper—he had been the referee *Nature* assigned to judge Stephen's work. Stephen appealed the decision, and the rejection was reversed by a second referee. The paper was published later that year.

Stephen wasn't daunted or insulted by this struggle, as some might have been. Jane noted that he took the struggle with Taylor "in a good-humored way," but that it also "served to reinforce his determination to fight against all odds, whether physical or in physics."

Those were the days when Stephen often didn't get the respect he deserved. He was easy to dismiss or underestimate. In Cambridge he didn't even have his own office—he had to share with another faculty member. True, he wasn't yet the superstar that Hawking radiation would eventually make him, but by then he *had* done several great pieces of work. Despite that, at a Cambridge dinner party a senior fellow spoke of him as if giving him even that spot in a shared office was doing him a favor. "As long as Stephen Hawking pulls his weight, he can stay at the university," the man said, "but as soon as he ceases to do that, he will have to go."

Once, before Stephen was well known, as he drove his wheelchair down a sidewalk in Pasadena, a passerby stopped him to hand him some cash. The passerby pitied Stephen and assumed, from his disability, that Stephen was indigent. Based on the way the man addressed him, it seemed he also thought Stephen must be mentally deficient. Before Stephen became a celebrity, strangers who saw him tended to judge him that way, to see him as damaged goods. Their reaction was intuitive, and not shaped by fact or nuance. He'd be treated as if being physically feeble meant that he must also be feeble mentally. That didn't insult him or make him angry. He just laughed it off.

In August 1974, just months after the Rutherford conference, Stephen traveled to Pasadena to begin a year at Caltech as a Sherman Fairchild Distinguished Scholar. It was this visit that launched his habit of visiting Caltech each year. He became a regular there, often staying more than a month. The trips were all arranged by his friend Kip Thorne.

Kip, who was about Stephen's age, studied classical general

relativity, the original Einstein theory without quantum modifications. At the other end of the Caltech theorist spectrum were two Nobel laureates who specialized in quantum theories and for the most part ignored general relativity. They were the two most influential theoretical physicists of their era, Murray Gell-Mann and Richard Feynman.

A decade after Stephen's Fairchild year, I arrived at Caltech and had the office next to Murray, and down the hall from Feynman. Gell-Mann was "Murray" to me and to most people. He derived his greatest fame from discovering a mathematical scheme to classify and understand the properties of elementary particles. The achievement earned him comparisons with Dmitri Mendeleev, who accomplished an analogous feat when he invented the periodic table of elements.

Feynman was "Dick" to a far smaller group. His most important contribution was to formulate a new way of conceptualizing quantum theory and of doing the calculations one needed to carry out in order to apply it—called Feynman diagrams. Like Stephen, he had invented his own way of picturing things and of doing the math, but unlike Stephen's, his had broad applicability in quantum physics. He published it, and it became a standard tool in elementary particle theory.

Murray and Feynman were friends and also rivals, but one thing they had in common was being rough on speakers whose theories they had issues with. They were both in the audience when, in the fall of 1974, Stephen gave his second important talk on black hole radiation, at the weekly Caltech physics colloquium. Stephen's student Bernard Carr had accompanied him to Pasadena and was again at the talk, showing the slides.

Everyone had heard of Stephen's theory by then, and this time the reception was polite. Murray didn't say much during the talk, but he didn't open a newspaper and start reading, as he sometimes did to show his lack of interest. Feynman would get up and walk out if he didn't like what he was hearing, but now he stuck around, asked questions, and expressed opinions. He even scribbled notes on the back of an envelope.

Among physicists Feynman was a living legend, and both Carr and Stephen were big fans. In the 1980s Feynman would gain fame in popular culture as well, after publishing a few best-selling books of anecdotes, and especially after his work on the presidential commission investigating the 1986 space shuttle *Challenger* explosion. On that commission he kept his distance from the government-affiliated panelists and became a severe critic of NASA's approach to safety issues, especially its tendency to underestimate risky flight conditions. Then he single-handedly identified the cause of the tragedy: they had launched the shuttle in dangerously cold weather. The result was a bad seal that developed when rubber joints called O-rings lost their flexibility. He demonstrated the effect on national television by dramatically dropping an O-ring into a glass of ice water and then pounding the ring on the table. It had become as rigid as a hammer.

Carr had been encouraged when he saw Feynman taking notes during the talk, but he noted that afterward, Feynman dropped the envelope in the trash. Carr was disappointed. The envelope's landing place didn't seem to signal immense interest in the topic. Still, Carr grabbed it as a souvenir. It had a dozen or so equations on it—and a sketch Feynman had made of Carr. Carr still has it.

Feynman dropped by Stephen's office after the lecture. Carr was there, too. Feynman said he had more questions about Stephen's work. He seemed skeptical. Stephen's distorted voice was very hard to understand, and there were no slides to illustrate what he was saying, so Carr had to translate.

A couple of days passed, and Feynman came by Stephen's office again. He had reproduced Stephen's discovery using his own Feynman diagrams. Now he believed Stephen. Stephen, still not an expert in quantum theory, wasn't very familiar with Feynman's methods, but they'd eventually become his favorite tools, and Feynman a friend.

In the weeks and months after Stephen's discovery, more and more theorists grew to accept it. Often, they would do as Feynman had—derive Stephen's result employing their own approach. Several published those alternate derivations. To this day there has been no experimental confirmation of Hawking radiation, but the fact that numerous physicists—each melding general relativity and quantum theory differently—had all come to the same result eventually led to its universal acceptance.

Ironically, the last holdout had been Zel'dovich, the man who had started Stephen down this road with his spinning black holes. His resistance ended one night in September 1975. Zel'dovich called Kip, who had been visiting Moscow but was packing to leave. He insisted that Kip come to his apartment. When Kip arrived, Zel'dovich was jubilant. After a year of trying, he had found an error in his calculations, and upon correcting it he, too, was led to Hawking radiation. He was beaming about it. There is joy in new understanding, even when it means that you were wrong.

Stephen was on one of those yearly pilgrimages to Caltech that he'd been making for decades, and we were using it to push our book forward. Feynman had died in 1988. Murray was near eighty, and in the 1990s had retired to the Santa Fe Institute in New Mexico. But Kip was still going strong, now running a group that did numerical relativity, the then-new field in which theorists "solve" the equations of general relativity by computer rather than through mathematical manipulation. The disadvantage of that approach is that the answers you get are just graphs or tables of numbers rather than the meaningful mathematical expressions you get in the traditional approach. The advantage is that a table of numbers is better than nothing, which is what you get when you try, and inevitably fail, to solve the equations mathematically. Kip's group focused on colliding black holes or black holes colliding with neutron stars. The aim was to provide precise descriptions of the gravitational waves such collisions would emit, for use by scientists at the LIGO gravitational wave observatory that he had cofounded in 1984.

I'd invited Stephen to my house for dinner after our first day of work, along with that evening's carer and Joan, who had also accompanied him to Pasadena. I felt bad for not inviting all Stephen's carers and the others who'd traveled with him since I knew them all well. But he had too big an entourage, so I had to limit the group.

Stephen's care, especially on trips, was a complicated endeavor. Even the night shift wasn't easy. The night work would start when Stephen was ready for bed. To communicate that, he'd say, "We'll go through now," meaning, "Let's go through to the other room to start preparing for bed." The abbreviated phrase arose because, if he was with guests, it was more discreet than saying, "I want to go to bed now." Once he'd uttered this, his carer would interrupt and tell the guest it was Stephen's bedtime as if it had been her idea. Sometimes Stephen would initiate this just to get rid of the guest, and after the person had left would rescind his request—he liked to do an hour of emails before retiring. Or, if he'd been out on the town, he'd want a bedtime snack. His favorite was poached eggs and mashed potatoes.

When Stephen really was ready for bed, the evening and overnight carers had to overlap for an hour or so. First, every night, there was the bath. He loved his bath. He liked it extra hot. They'd undress him and then, when at home, fit a sling over him so they could move him to the bath using a hoist on the ceiling that had tracks on it. On the road they'd have to lift him. While he was soaking they'd heat up some towels and fill the wheelchair with them. When they got him out, they'd wrap him in more towels, and transfer him back to the chair. Then they'd get his "nebulizer" going to humidify the air, change his tracheostomy tube, dress him in his nightshirt, and put him in bed. In these later years he'd be on a ventilator all night, so they'd have to hook him up to that, too.

Once in bed Stephen was in his most vulnerable state because communication was even tougher than usual. If he wanted to say something that he couldn't get across with his face, his carer would start pointing to the letters on the spelling card. Stephen's carers had to watch him closely. Whenever he awakened, they'd try to figure out if he needed something. A dozen times each night he'd ask, with his eyes, to be turned and for his pillows repositioned. He couldn't shift his weight from time to time as healthy people do, so he'd get uncomfortable. His bones hurt. In addition, his carers would have to listen to make sure his stoma was clear so he could breathe. And every couple of hours, as he slept, they gathered his vitamins, mixed them with liquid, and fed them directly into his stomach through the peg. "What you do for a newborn baby you did for Stephen. All of us," said his carer Viv. "When I came off shift and he was still alive, I felt I'd done a good shift. Because he was alive. Because I'd kept him alive."

This evening most of the other guests were already there when Stephen and company showed up in a rented van specially outfitted for the disabled. My house, however, was not specially outfitted. I knew that with the help of a couple of guests I'd be able to carry him, in his chair, up the five or six steps to my front door. Not that it would be easy. He had a heavy-duty wheelchair with a motor, a battery, and a computer mounted on it. Plus, of course, there was Stephen. He was short and had always been slight, and on that evening he weighed about ninety pounds. We could have managed it, but I knew he was fed up with buildings that had stairs and no ramp, so I'd made a ramp, a slab of wood I had cut to throw over the stairs.

Stephen would have been fine with the carry-up-the-stairs plan at a private gathering, but in public places he had little patience for facilities that hadn't been made accessible. He knew that, compared to most other disabled people, he had it good. Thus if he had difficulties, those of others would be worse, and it angered him when accommodations weren't made.

One time Viv took him to see his mother in Stratford-on-Avon. They went to a restaurant in a National Trust property nearby, a historic old building. Stephen needed to urinate, so he told Viv he "needed the bottle." That was their code for that request—the bottle being a plastic urinal that he would pee into. But first they needed to get to a bathroom, and the restaurant had no disabled toilet. Viv was pondering where else they might go when Stephen asked to be wheeled around to the back of the kitchen. This puzzled Viv. She told him that it didn't seem like a good idea, but he repeated his request. She complied. It was faster than having a discussion about it.

When they got there, Stephen said, "I need the bottle."

Viv started to wheel him away, to find a place where they could have some privacy. But as she started to push the chair Stephen made a few angry contortions of his mouth and nose. Viv stopped.

"I need to stay here," Stephen said, with the volume turned up high.

"You can't do it here!" Viv said. She must also have been loud about it because now the chef came out.

"What's going on?" he asked.

"Disabled toilet," Stephen said, still loud.

"Sorry, we haven't got one," the chef said. His face said, *What made them think they'd find one back here?* He shook his head and retreated into the kitchen.

"I need the bottle," Stephen said, glaring at Viv.

Viv wheeled him into the hedges outside the kitchen's back door and pushed him in as far as she could get him. Stephen did his thing. She wheeled him out of the greenery.

"Empty it now," Stephen said.

"I can't!" Viv said. "The kitchen is right here!"

"Empty it," Stephen said.

Viv meekly started to empty the bottle into the dirt under the hedges. Just then the chef came out again. When he realized what they'd been doing he hit the roof. In the midst of his tirade came words from Stephen.

"Disabled toilets," Stephen said, with the volume still on high. He flashed his angry grimace as the words came out. This was Stephen hitting the roof.

The chef was taken aback, but Viv wasn't about to stick around to extend the altercation. She hurriedly wheeled Stephen away.

The incident left Viv embarrassed. She got embarrassed a lot, caring for Stephen. A year or so later Stephen wanted to visit that restaurant again. Viv took him back to it. They had added a disabled toilet.

Stephen appeared touched that I'd made the house accessible for him. He didn't know I'd heard the story of what happens when you don't. He stayed at my party for quite a while and seemed to be having a good time. My friends were a bit in awe of him, except for one who had a Nobel Prize and knew him. At Caltech in those days, every other faculty member seemed to have a Nobel. Since there are an infinite number of parallel universes, I figured maybe I'd get one, too, but mine would be in one of those.

People often wondered why Stephen didn't get a Nobel for his discovery of Hawking radiation. One of my friends at the party apparently wondered that, too. If he'd asked me before the party I'd have explained. Also, if he'd asked me after the party. Instead, he asked during the party. And he didn't ask *me,* he asked Stephen.

That made me uncomfortable, but it didn't faze Stephen. It was late and Stephen was tired by then, and it took a while to type it out. During the interval we'd moved on to talking about banjos, which the Nobel laureate was obsessed with. This surprised my other guests. They'd expected him to be an expert on string theory, but not that kind. Then, finally, came Stephen's words, accompanied by a matter-of-fact expression on his face. "Need to observe it," he said. We all thought back and realized what he was referring to. I filled in the details for my friend. The radiation hadn't been observed—hence no Nobel.

There is a lot to object to in the awarding of Nobel Prizes. The times they gave it to the wrong person or overlooked the right one. The times they gave it for something that didn't warrant it or overlooked an advance that did. But one consistent principle the Nobel committee seems to adhere to is that a theoretical advance will not bring a prize unless it is confirmed by observation or experiment (and even then, to the annoyance of us theorists, they often award it to the people who did the experiment rather than the theorist upon whose breakthrough it all depended).

Unfortunately for Stephen, there are many obstacles to observing Hawking radiation. For example, before you can observe a black hole radiating, you have to locate a black hole. The first object widely accepted to be one, as I mentioned, was an object called Cygnus X-1. It had taken about twenty years of experimental work by hundreds of scientists from the early 1970s through the early 1990s to achieve that confidence. During that time, Stephen bet Kip that it would turn out not to be one. His reasoning was that he wanted it to be one, and by betting against his wishes he'd have a win either way. Since then we've located many black holes. In fact there seems to be one at the center of nearly every large galaxy.

But there's another problem—the "Hawking temperature" of a black hole is typically less than a millionth of a degree. That's very close to absolute zero, too close to be detectable with today's technology. Also, a typical time for the black hole to lose an appreciable amount of mass is around 10^{67} years, an unimaginably long time (the present age of the universe is about 10^{10} years). So we're not going to be able to measure any of them shrinking.

Indirect evidence supporting Hawking radiation did finally come in 2019, a little more than a year after Stephen's death, in an experiment performed by a team of physicists at the Technion, in Israel. The scientists started with a "sonic analog" of a black hole. Think of a fluid that is flowing faster than the speed of sound.* Due to the speed of that flow, if there is a source of sound in the stream, you'd expect that no sound waves will escape in the backward direction—they cannot outrun the flow of the stream itself. That's analogous to the black hole property of allowing no signal, such as photons of light, to escape its gravity. The question the scientists asked was, in this model would they observe the analog of Hawking radiation?

In the Israeli experiment, the fluid consisted of 100,000 ultracold rubidium atoms. They played the role of the black hole. Quantum particles of sound, called phonons, played the

---

* Sound moves through different substances at different rates. The speed of sound I'm referring to here is the speed of sound in that particular fluid.

role of photons, particles of light. The scientists found that some phonons did indeed emerge from that sonic black hole, and the characteristics of that escaping energy matched what Stephen had predicted. It was the phonon analog of Hawking radiation. Had Stephen survived, that might have been enough for the Nobel committee, but another thing the Nobel committee does not do is make posthumous awards.

Hawking radiation was important because it represented the first important instance in which general relativity and quantum theory were applied to the same system. We still have no complete theory of quantum gravity, but by having black holes as a mathematical laboratory for mixing general relativity and quantum theory, physicists have been able to learn something about the properties and principles of that elusive theory.

8

The morning after the party I was in an impatient mood. We'd made no concrete progress the day before. We'd sat side-by-side, I staring at Stephen, he staring down at his computer screen, typing out long (for him) commentaries about the book and what he wanted it to say. He was questioning its fundamental thrust. This despite the months we'd spent developing the very concrete plan that we'd agreed upon—and despite the fact that we'd already written five of the eight chapters. I added my two cents now and then, and he replied. We traded ideas, and some were good ones. But why the second-guessing? Was it procrastination, a way to avoid having to write? I thought not. Stephen's psyche appeared as healthy as ever, but for some reason our book was having a midlife crisis.

We'd agreed that morning to meet at the Athenaeum—the Caltech faculty club—for lunch before going to Stephen's office. Stephen showed up about a half hour late. Truth is, on this occasion his tardiness was really my fault, for having kept him out so late at my party. But he'd often start later than he'd said he would. I understood that. It took a lot of prep in the mornings to get him ready. Things could drag out. A call would have been nice—to warn me—but such calls never came. He was like the trains running late. You can take it or leave it but you can't change it. People called it Hawking time.

Locals admire the Athenaeum. Red tile roof, lots of archways, elements of Spanish and Italian villas. Mediterranean revival. The dining room has twenty-foot gilded ceilings with full-height windows, ornate chandeliers, lots of dark wood, and, my favorite, oil paintings of long-dead science greats. The "Ath" is considered a venerable place. But like space-time, what is venerable depends on the observer. Cambridge traced its founding to a 1231 charter from King Henry III. There, even a recent addition would be centuries older than anything at Caltech. So to Stephen, the reverence afforded the Ath must have caused amusement. Then again, Cambridge didn't have warm sunny winters and ubiquitous wheelchair access.

Though the atmosphere at the Ath couldn't stand up to that at Caius, the food could, and like an army, Stephen marched on his stomach. Today they served roast beef. That made Stephen feel at home. We were at the tail end of the meal, which meant that everyone but Stephen was already finished.

I was a bit nervous because I'd spent the morning contemplating how the previous day had gone. It had been our first workday since his arrival, and upon reflection I became disturbed by our interaction. It wasn't just that we were revisiting decisions we'd already made; he also didn't seem to have done any of the work he'd promised to do—over several months—before coming to Pasadena. Back when I was just starting out and Feynman was dying of cancer, Feynman had told me that examining a relationship, or your life, has its uses, but if you're happy it's best to avoid it. But I wasn't happy, and although I'm not big on relationship talks, I felt I needed to have one with Stephen.

A personal discussion was not something I wanted to engage

in at lunch or in the company of others, so I was holding off. Of course, with Stephen you were always in the company of others, or almost always. Here at lunch there were two carers, David and Mary. Mary was the carer on duty. At the office David wouldn't be with us, and I figured I could wait for Mary to take a bathroom break. I was biding my time, waiting for lunch to be over. But Stephen was eating even more slowly than usual.

"Come on, have another bite," said Mary.

Stephen curled his mouth downward, *no*.

"Oh, come on! It's soooo good. Just open up and have one more. You can do it! You've hardly eaten." She put a hand on his hand and spoken to him with the slow, overenthusiastic cadence you'd use when addressing a baby. Stephen had to be cared for as if he were a baby, but he didn't have to be spoken to that way. Still, some carers did.

The carers each had their style. They all seemed to love him, but each in their own way. Some were serious matronly types. Others flirted with him. They'd wear tight, low-cut tops, and when they leaned over to adjust something on his person, they made sure he got a dose of their bosom. Mary wasn't like that. She had a matronly look and an approach to match. She was a babyer, and she wasn't the only one. It seemed to work, because Stephen opened his mouth. She fed him the spoonful, then wiped his chin with a napkin. Stephen didn't discourage either the flirts or the babyers. They knew better than to treat him that way with strangers around, but when alone or among friends, he seemed to enjoy the attention of both.

That Stephen attracted nurturing was easy to understand. But he also attracted affection. I felt that for him almost from the start. Part of it was his eyes, blue and full of character. They could impart great warmth. They could to speak to you. They could make you feel connected. To those who were his friends, they were affectionate. To those who didn't know him, they were inviting. To those who were annoyed with him, they were disarming. When he was in pain he'd scrunch them up, and you'd feel that, too. And if you made him angry, his eyes made you wish you hadn't.

Waiting for Stephen to finish with his food, I'd started talking to David. We'd somehow ended up challenging each other in arm wrestling. This had been going on for a few minutes with no winner but a lot of grunting. Stephen watched as Mary flagged someone down to remove the dishes.

"What do you have for dessert?" she asked the student-busboy-waiter. "It will have to be gluten free."

Stephen always ate gluten free, though on occasion, when it was convenient and not obvious, I'd seen carers feed him food that wasn't. That never seemed to cause any ill effect. On the other hand, Stephen had told at least one friend that he wasn't really allergic to gluten. So why did he allow, or require, his carers to restrict his diet like that? It was all a mystery to me, but I never asked. In this instance it didn't matter, because Stephen showed with his grimace that he didn't want any dessert. Then he went back to watching us arm wrestle.

I wondered if Stephen found our behavior childish or irritating. It wasn't exactly dignified. But he didn't seem to mind. He took pleasure sometimes in watching others do what he couldn't. I noticed that at bars, if young people were dancing, he liked to watch. But I guess he had finally had his fill of vicarious wrestling because at this point he said, "Let's go."

I girded myself. We'd soon be at his office, and I'd decided that the minute Mary stepped away to powder her nose, I'd bring up my issue. Stephen and I had debated many things, but this one wouldn't be an intellectual discussion. I wondered what Stephen's style was when talking about personal issues. He was, after all, a species of rock star, and most people probably kept any displeasure with him to themselves. The thought of his slow communication didn't make it any easier. Waiting seven minutes for a few sentences about the quantum origin of the universe was one thing, but this discussion had the potential to get uncomfortable. Who wants to have an uncomfortable talk in slow motion?

I'd heard once that a good relationship is not one that is conflict-free, but one in which conflicts are handled with mutual love or at least respect. If your significant other makes the morning coffee weaker than suits you, you're supposed to say, *Hey, honey, would you mind throwing in a few extra grounds next time, to hit that sweet spot where we both like it?* That's supposed to elicit a response like, *Of course, dear!* I'd had relationships, though, in which the exchanges were more like, *Hey, the coffee's too weak,* and the response is, *Next time make it yourself!* What type of relationship would Stephen's and mine be? I figured that what I had to say, and how he answered, could be a test of things to come. It could be a way to deepen our connection or, if it went south, it could make our future work awkward for us.

Mary was dabbing at Stephen's mouth, and I had already popped up out of my seat when, seemingly from nowhere, Murray Gell-Mann walked up. Apparently he was visiting from New Mexico. I'd seen him only every now and then since he left Caltech in the 1990s. I thought about how, each time I saw him, his hair had grown just a bit whiter, his posture more slumped. Though still in his late seventies, he seemed a little less sharp, too—not that I would've wanted to challenge him to a physics contest. In comparison to Murray, Stephen appeared ageless. In the years I was friends with him, I didn't see Stephen decline appreciably, especially not in the intellectual domain. It was only his communication speed that diminished—and his reading ability, which slowed due to his eye control issues.

I greeted Murray, who then turned to Stephen with a big smile. He was standing off to Stephen's side. "Hello, Stephen!"

he said with enthusiasm. "So good to see you!" Stephen didn't say anything but aimed his eyes at Murray, put on a huge smile, and held it for a few moments.

"I won't take up your time," Murray said. "I just saw you were here and wanted to say hello. You're looking good!" Stephen then flashed another smile, this a shorter one, but his eyes showed his appreciation.

With that, Murray walked off. Stephen owed Murray a lot. Back in August 1985, just after Stephen's tracheostomy, it became clear that it would be difficult for Stephen to survive without round-the-clock nursing care, which Britain's National Health Service wasn't going to pay for and which he couldn't afford. Kip suggested that Stephen apply to the MacArthur Foundation for financial support. Murray, then still at Caltech, was on the foundation's board. And so began a generous string of grants that enabled Stephen to hire carers in those days before his books hit it big.

These weren't the sort of grants the foundation usually awarded. MacArthur is famous for its "genius grants"—one-time awards meant to be given to needy, unrecognized, and promising young individuals in various arts and sciences. In practice, the genius grants are all too often given to those who are already famous and well off. Stephen, then forty-three, was as much a genius as any of those genius grant recipients but not yet a celebrity—he'd just started to write *A Brief History*—and he really did need the money.

In physics, you don't get paid for what you discover. You publish your work, and in exchange you get a tenured position and the satisfaction of having figured something out. You are supposed to be content with a modest salary and a secure job that pays you to do what you love. In 1985 Stephen's annual salary was about $25,000. Those earnings don't go far if you have ALS. Fortunately, Stephen's work on Hawking radiation had elevated him in the physics community by then, if not in the world at large. His name had already been on the map, but after Hawking radiation it was written in the largest of fonts, and so the MacArthur Foundation was happy to help and gave the money to Cambridge University to administer.

Stephen and Jane were grateful for that and felt bad for those who had ALS but hadn't discovered Hawking radiation. Those individuals would be stuck with what the health service in their country provided—a bed in a nursing home where they would lie incommunicado and isolated, relatively unattended and with little stimulation. In that setting, any of the close calls Stephen would suffer in the coming years would probably be fatal. Without his round-the-clock care, Stephen once said, "I would last exactly five days and die."

The carers who watched over Stephen during the years I knew him were not nurses, but those who looked after him in the early days were. One of the applicants for the positions the 1985 MacArthur grant funded was Elaine Mason, a tall woman with long, wavy, red hair. She was a nurse at Addenbrooke's Hospital at the time but preferred to work with a single chronically ill patient. She had a long history of caring for others, including four years tending to the wounded after the 1971 war in Bangladesh. She got the job.

Of the different approaches to caring for Stephen that I mentioned, Elaine would use them all. As the serious nurse she would in coming years save his life on more than one occasion. She would also baby him at times, but in a playful way. And she definitely knew how to flirt. Elaine quickly became Stephen's favorite carer. Stephen was then in his early forties. Elaine was a thirtysomething who rode skateboards. Legend has it that while Stephen was receiving an honorary doctorate from Harvard, she got bored and started doing cartwheels. If Stephen vicariously enjoyed physical play, Elaine was a great person to enjoy it through. Maybe one reason they bonded was that she had the flamboyance he would have exhibited if he'd had the use of his body.

For her part, Elaine wasn't put off by Stephen's physical condition. Just the opposite: she was drawn to it. Her first husband, David Mason, said that all Elaine really wanted was someone who needed her. Unlike Jane, Elaine would accompany Stephen on most of his trips abroad. She loved that he'd travel, work on physics, write books, or just talk, undeterred by the enormous effort it took. She loved his strength. She'd speak and listen to him patiently, appreciative of the time and energy he expended to communicate with her, and she began to confide in him.

Meanwhile, eight years earlier, his wife, Jane, had met someone else to talk to and confide in. His name was Jonathan Hellyer Jones, the choirmaster at her local church. Stephen was physically capable of having sex, but at some point that had stopped happening with Jane. His condition meant that Stephen had always been a completely passive sex partner as well as a fragile one. Over time, his fragility caused Jane to worry that sexual activity might kill him. Making love to him became a frightening and empty experience. Even the thought of sex with him felt unnatural, and her desire for him faded. He had the needs of an infant and "the body of a holocaust victim," she said. Their passion for each other extinguished, Jane's marital relationship with Stephen devolved into that of a carer—feeding him, bathing him, brushing his teeth, combing his hair, dressing him. With Stephen submerged in his work and Jane tending to all his practical needs but feeling taken for granted, she and Hellyer Jones became lovers.

Jane talked to Stephen about her affair and got his blessing. Her idea was that she and her lover would keep their relationship private and discreet. Their family would evolve to include them all. It would be a "new arrangement," a kind of extended family. What Jane didn't expect was that Stephen would extend the family again—to include Elaine.

If Stephen and Jane had followed a path from man-and-lover to infant-and-carer, he and Elaine followed the reverse trajectory. This led to a new "new arrangement." It was a constellation as complex as any in the night sky, encompassing Stephen, Elaine, Jane, and Jonathan; the three Hawking children; and their various interconnecting relationships. There was also an added complication: Elaine was still married to David Mason.

That Stephen would mirror Jane and take a lover hadn't pleased her. Jane knew about symmetries from Stephen's work, but she didn't like this one. Still, for a while they all stayed committed to the one-big-happy-family approach. That "while" lasted about as long as you'd expect. Stephen moved out to live with Elaine in 1990, and a decade later he and Elaine built the house that Stephen would live in for the rest of his life. He married Elaine in 1995, shortly after she divorced David and he divorced Jane. He was then forty-eight.

After they were married, Elaine shed the official role of nurse in Stephen's life. She wanted to help him, to enable him to do all the things that he wanted to do, and to care for him, but as a wife and not a carer. She didn't want to be the one who'd chop up his meat and spoon-feed him, but she did love cooking for him. She prepared curries, roasts, piles of fruit, rollmops, all his favorite foods. If he wanted something in particular for dinner and they didn't have the ingredients, she'd run to the store and get them. She loved going out with him, too. Sometimes when they were going somewhere special, she'd get a new dress, and when he came home at night, she'd run to greet him. "I've got to show you, Stephen!" she'd say, and she'd go upstairs, change, and then do a fashion show. She loved to hold his hand, and he loved the affection and returned it.

Though Elaine couldn't sleep in Stephen's bed, sometimes she'd come down in the middle of the night just to look in on him, or to sit with him and touch him. She felt that Stephen was her gift. "I helped Stephen but he helped me," she told me. "I came from a dysfunctional family. My parents didn't look after us very much." She wasn't in love with David, she said. "I loved him, but we weren't *in love*. I married him because I was twenty-five and he was the first man who asked me, and that's what you did. So the feeling of being loved was special. And I was in love with Stephen and he was in love with me. He accepted me and loved me for who I am inside."

Stephen had started the 1960s as a lackadaisical undergraduate and ended the 1970s as a dominant figure in the fields of quantum gravity and cosmology. He never accepted the idea that he was another Einstein, although there were some similarities in their approach, both to physics and to life. They were both geniuses, both mavericks, both visionaries, both distinctively talented at seeing through the clutter to identify the important questions. But they lived in different times, times in which physics was asking different questions, and they faced different personal challenges. That makes it difficult to compare their talents. It's not difficult, though, to see that they influenced physics on different scales.

Einstein made broad and revolutionary contributions on several fronts. In addition to his special and general theories of relativity, he achieved what many considered to be the first proof of the existence of atoms, and he was the first to recognize Max Planck's quantum hypothesis as a universal truth of nature and apply it outside the narrow realm in which it was discovered. He wasn't merely a dominant figure in one area of physics; he reshaped the foundations of the entire discipline. That, Stephen did not do.

The influence of Stephen's work, on black holes and the origin of the universe, was for the most part limited to the fields of cosmology, general relativity, and the search for a theory of quantum gravity. By head count, at least, this was only a small segment of the physics community. Of course, it is impossible to know what Stephen could have accomplished had he lived in a different era, and/or had he been healthy. Regarding the latter possibility, however, Stephen himself seemed to believe he might have accomplished less, because he would have been less focused, less driven by an awareness of imminent mortality.

While most physicists considered Stephen's work on black hole radiation to be his greatest discovery, Stephen was not among them. In his view his most important contribution was a far less influential effort, his research in the 1980s on the quantum origins of the universe—a theory he dubbed the "no-boundary proposal." It was so esoteric that once, after Stephen had given a talk about it to an audience of hundreds of physicists, a colleague remarked that "maybe twenty people in the audience really understood his lecture. It was really heavy stuff."

At first I was surprised at Stephen's assessment. But when I thought back to the reasons he got into physics in the first place, his judgment made sense. The Holy Grail for Stephen was to understand the beginning of the universe. He wanted to know where we all came from. And that's what he believed he accomplished with the no-boundary proposal.

The no-boundary proposal was a natural outgrowth of Stephen's earlier research, the culmination of the work he had by then been doing for twenty years. His first two research programs—on the origin of the universe and the laws of black holes—were based solely on general relativity, without taking into account the principles of quantum theory. After Stephen studied the quantum literature, he applied what he'd learned to black holes, revised his earlier ideas, and discovered Hawking radiation. Now armed with quantum theory, in his next major research effort he would revisit the origin of the universe. That culminated in his no-boundary proposal, research that he conducted with his friend Jim Hartle of UC Santa Barbara, a couple of hours north of Caltech.

The no-boundary proposal was based on a seemingly odd idea. Quantum theory is normally considered to be, as I have described it, a theory of the small. It is usually used to portray a system comprised of an atom or molecule, a subatomic particle, or a compact collection of such objects. If it is to be applied to the entire universe, one would therefore think that its relevance would be confined to the early universe, when the entire cosmos was atom-sized. But Stephen had a greater ambition, to treat the universe as a self-contained quantum system throughout its history, from its microscopic early days to its vast present-day existence. His main tool in that effort was the revolutionary approach to quantum theory that had brought Feynman the Nobel Prize in 1965.

In the original conception, quantum theories describe the state of a system by a certain mathematical construct—a wave function. The wave function contains everything one can know about the system. That information allows you to calculate various probabilities—for example, the chances that, if you make a measurement, the particle will be found to have a certain position, momentum, or energy. Quantum theory dictates that this is the best you can do—you cannot, as you could in Newtonian theory, *guarantee* what the result of a measurement will be.

If that were all there was to it, the wave function would be like a reference manual describing the system at some given time. But systems change over time, and the wave function changes to reflect that: given the wave function at one moment, the mathematics of quantum theory tell you how to determine the wave function at any other time. That's a vital aspect of quantum theories because the usual question in physics is, given a system that starts out in an "initial state," what are the chances that it will evolve to various possible "final states" at some later time?

The scheme I've just described was employed with great success to explain the properties of atoms and the chemical elements from which they are made. Other quantum theories—quantum field theories—were then developed to describe the interaction of elementary particles. For example, the electron, positron, and photon are described by a field theory called quantum electrodynamics. To carry out calculations in theories like quantum electrodynamics was exceedingly difficult. That's when, out of the blue in the late 1940s, Feynman formulated his new approach to quantum theories. It looked nothing like the original scheme.

In Feynman's approach to quantum theories, wave functions are not fundamental. Instead, to find the probability of a particular final state of a system, you start with its initial state and consider all the possible ways, or histories, through which it could have developed into the final state. You then add up the contributions from each history, using certain rules Feynman developed. The method is sometimes called the Feynman sum over histories.

To illustrate the idea, suppose you are trying to calculate the probability that a quantum particle evolves from an initial state in which it is located in a lab at Caltech to a final state in which, at some later time, it strikes a detector in a lab on the moon. In Feynman's formulation, you do that by including contributions from all possible paths between those two labs. That would include paths in which, along the way, the particle goes past Jupiter or circles the earth a million times. It even includes paths that violate the laws of physics, in which the particle flies all over the universe, traveling faster than the speed of light or moving backward in time. Most paths are of that nature. But Feynman's rules dictate that the direct "straight line path" contributes the most, while the "absurd" paths contribute very little. Still, there is an endless set of paths, each of which does contribute something, however great or small.[*]

Stephen no doubt admired Feynman for the elegance of his ideas, but I think he also felt a kinship with him as a fellow maverick who shook things up and had to fight to convince others of his ideas. When Feynman unveiled his new approach at a conference in 1948, for example, he met the same sort of resistance that Stephen faced when he announced Hawking radiation. Prominent physicists like Niels Bohr, Edward Teller, and Paul Dirac all said Feynman's method was nonsense.

Feynman's theory was indeed radical, and could at first glance appear outrageous—with its particle paths that zigzag all over the universe. In his derivation of it, Feynman had, like Stephen, cut corners and deviated from mathematical rigor. For example, the way you are supposed to sum over paths seemed to

---

[*] Everyday (macroscopic) objects are a composite of a great many molecules. In such objects, the contributions from most of the paths cancel each other out, creating something that, when viewed as a whole, obeys Newton's laws. In physicist's language, one says there is decoherence by coupling to internal degrees of freedom. See Todd A. Brun and Leonard Mlodinow, Decoherence by coupling to internal vibrational modes, *Physical Review A* 94 (2016).

violate certain fundamental mathematical principles, but Feynman paid no heed to that problem. Also like Stephen, Feynman preferred to think in terms of pictures instead of equations, an unfamiliar approach that added to the skepticism. "It seemed like a sort of magic," said physicist Freeman Dyson.

But Dyson and others eventually showed that Feynman's method could be given a firm mathematical footing, and that—though the theory provided a different picture of what is going on—it would always make the same predictions for the outcome of experiments as the earlier formulations of quantum theories. Feynman was not proposing any new laws of quantum physics. Rather, he was offering a new way of *looking at* quantum physics, a new way of thinking about the quantum universe, which led to some amazing new insights.

In certain fields, such as elementary particle physics, both Feynman's conceptual picture and his methods of calculating the predictions of the theory proved far superior to the old ones. As a result, today Feynman's approach is a standard tool in theoretical physics. Stephen studied the method during his year as Fairchild Fellow at Caltech and while there had the opportunity to learn about the approach from its creator himself. That's how, ten years later, Stephen came to employ it in his no-boundary proposal—the only difference being (and it is a massive difference) that since Stephen was seeking to trace the quantum history not of electrons or photons but of the cosmos, in his work the entire universe would play the role of the particle.

To apply quantum theory to the entire universe raises many questions. For example, when theorists use the Feynman sum over histories to analyze the motion of an elementary particle, the information they are starting with, and which they want to trace to a later time, has to do with observable attributes such as position. But the universe has no "position"—the universe is all there is.

Instead of concerning itself with position or any of the other variables of interest to particle theorists, Stephen's theory revolves around variables pertaining to the geometry of space-time—to its curvature, as defined at every point. What does that mean? Consider the space we live in. It has three dimensions—at any point on earth, for example, you can move north/south, east/west, up/down, or in any combination. Mathematics provides a way to describe that three-dimensional space, and, indeed, a space of any number of dimensions. It also supplies a definition of what we physicists mean when we say that space is curved, as opposed to flat.

Since it is difficult to imagine the curvature of three-dimensional space, let's drop the up/down dimension and think about a world with just the north/south and east/west directions. That's a two-dimensional space. If you imagine those two directions being defined on a *plane,* that's a flat two-dimensional space. It is the type of space you learn about in high school geometry. It obeys rules such as the one declaring that the angles in a triangle sum to 180 degrees.

If instead you imagine the directions north/south and east/west as referring to the surface of a *globe,* that represents a curved two-dimensional space—a mathematician would say it has positive curvature. The surface of a *saddle,* in contrast, represents a space that is curved in a fundamentally different way—negatively.

Spaces with positive or negative curvature obey geometric laws that are different from the ones you learned in high school. For example, the sum of the angles in a triangle in a space with positive curvature is always greater than 180 degrees and in a negatively curved space it is less than 180. Such differences allow physicists to determine the curvature of the actual three-dimensional space we live in.

In general, a space can be positively curved at some points and negatively curved at others, as if tiny sections of globes and saddles had been cut out and smoothly patched together. And the magnitude of the curvature, whether positive or negative, can also vary. Space can be slightly curved at some points and severely warped at others, like the peaks and valleys of the earth's landscape. That's what's meant by "the curvature of space, defined at every point." It is that "landscape," and its evolution over time, that Stephen focused on in the no-boundary proposal. His theory wasn't meant to be a detailed theory of all the energy and matter that populate the cosmos—of the stars and particles and planets and people—but rather of the shape of physical space itself.

In calculating the shape of the universe over time, just as in calculating the evolution of a particle, one normally starts with the initial state. Nobody knows the initial state of the universe, so Stephen and Jim Hartle had to make a conjecture. Their guess was that when you go back far enough in time, the great compression of matter and energy into a small space changes things in a fundamental way—it makes time so warped that it is unrecognizable and essentially becomes another dimension of space.

It was Stephen who'd shown, in his Ph.D. thesis, that the "classical" big bang theory, based on general relativity, must have a singularity—a time at which various quantities such as curvature become infinite. Now, when he and Jim Hartle modeled the quantum history of the universe in the above manner, they found that the singularity Stephen had predicted would occur at the beginning of time was no longer there. The laws of quantum theory had inspired Stephen to revise his original theory of black holes and led him to Hawking radiation; now those laws were demanding a change in the scenario he'd put forth for the origin of the universe.

Stephen liked to use a metaphor to explain the new theory. Suppose you are somewhere on a straight train track that starts and ends at some point. Suppose also that moving back toward the starting point represents moving backward in time. In that picture, from wherever you are, if you start moving back in time, you'll eventually get to a point where time begins—you'll run out of track. That represents the singularity that Stephen described in his Ph.D. dissertation. But when you take quantum theory into account, Stephen said, the flat track looks more like a track on the surface of a globe, with south representing going backward in time and north going forward. Now suppose you start to move back in time—to move in a straight line, due south. In this situation you will never experience a point where time begins. There is no "boundary" to time, no beginning and hence no singularity.

That is the picture that Stephen derived from his ideas. It answered the question he had asked when he first got into physics: How does the universe begin? His answer was a surprising one—that there was, in the sense I've explained, no beginning, because time had turned into space.

To Stephen the no-boundary theory was momentous—not only due to the question it answered but also because of a question it raised. As he wrote in *A Brief History of Time,* "So long as the universe had a beginning, we could suppose it had a creator. But if the universe is really self-contained, having no boundaries or edge, it would have neither beginning nor end: it would simply be. What place, then, for a creator?" It was a question we returned to in *The Grand Design.*

Stephen's office at Caltech was pretty bare-bones. Off-white walls, metal desk, small window, one of those modest spaces reserved for short-term visitors. My office, which was in another building, was more comfortable and cheerful, but when Stephen was in town I'd spend more time in his than mine. I'd sit with him as we worked together, and I'd usually also get there before he arrived and stay after he left.

In the time since I'd last seen Stephen, his communication with me had been sparse. It had bothered me that he hadn't been more responsive, but I knew that for him writing emails was a slow process that he avoided unless absolutely necessary. I'd assumed that he was writing the sections we'd agreed he'd write and reading the ones I sent him. We'd planned to go over all that during his current visit, and then to push on from there.

That's why I was so surprised when, the prior day, he had immediately started talking about "global issues," about what we were trying to say in the book—topics I'd thought we'd already settled. I'd become increasingly uncomfortable as the day wore on and I gradually came to the realization that, in the time since my last visit to Cambridge, he'd hardly thought about the book. We had a deadline that I knew we wouldn't be held to—we were already far behind. But if, moving forward, Stephen would only work on the book when I was standing over him, I figured one or the other of us would be dead before we finished.

It was only a few minutes after we arrived at Stephen's office that his carer Mary left the room. As soon as she was out of earshot I started the dreaded conversation. I tried to sound casual. "That was a nice lunch, huh? The Athenaeum does a good job. By the way, I assume you've read over what I've written since I was last at Cambridge."

Stephen raised his brow, meaning yes, and he smiled. So he *had* worked on it. I was relieved. I was glad I hadn't come on stronger about this. How foolish of me to think he'd ignore his commitments. He started to type. After a minute or so, his voice said, "Yes, it was a good lunch."

I tried not to look the way I felt. "And our book?" I asked.

He started typing again. "I have been busy," he said.

"Did you read any of the pages I sent you?"

He grimaced *no*.

"Did you write anything yourself?"

Another grimace, another *no*.

I didn't know what to say. I'd put a lot of work into what I'd written despite having a class to teach, physics research of my own that I was trying to push forward, and another book—*Subliminal*—that I was in the midst of writing. I knew that he had even more on his plate, plus the enormous difficulty, for him, of doing *anything*. I thought I should feel sympathetic, yet I found myself feeling angry instead. Fortunately the response I blurted out unthinkingly was somewhere in the middle.

"I'm disappointed by your inactivity," I said. Having said it, I felt disappointed in me. Who was I to talk to Stephen Hawking that way?

He made a face. I tried to read his expression. What was he thinking? It wasn't an angry face. It looked like the face a dog might make after you've kicked it. Was he hurt by what I said? Was he sorry?

"I don't want to work on the book if you're not going to," I said. "We have to work on it together." He raised his eyebrows *yes* in acknowledgment. It seemed to be a friendly *yes*. That felt good.

I slid my chair a bit closer to him and took his hand in mine. It was warm and limp. I'd reached for it on an impulse. He seemed to be fine with it. I had the feeling he liked it. Maybe he cracked a little smile, or maybe that was wishful thinking on my part. We locked eyes for a moment.

"Do I need to go live in Cambridge until we get back on track?" I said, in a softer tone.

He immediately grimaced *no*. He definitely did not like that idea. I didn't know whether this meant he couldn't afford that much time or that he was repulsed by the thought of having me around that much. Either way, it felt like rejection.

I gave him his hand back. He started to twitch his cheek. He was typing.

"I admit I have been inactive," he said. "I couldn't get fired up about the book."

It felt bad to hear that from him.

"If it bores you, maybe we shouldn't do it," I said.

He indicated *no*.

"After our work yesterday, I am excited," he said. "I feel I now know where the book should go. I will be more active in the future."

Mary came back in the middle of that but sat in a chair and ignored us.

Stephen spent the next hour explaining his rethinking of the chapter outline we'd developed. Although the first five chapters hadn't changed much in this new version, he described them in detail anyway. Then he described the big changes he had in mind for chapters 6 and 7, which was what we were currently working on. When he was finished explaining, he said, "That's the first seven chapters. Let's concentrate on them for now." Our original plan had had eight chapters, and given that we were altering chapters 6 and 7, I expected that the final chapter would have to be altered as well. Based on what he said, I supposed that when we got there we would just wing it.

In the coming years I'd have many frustrations with Stephen. There'd be the interruptions, the starting times that turned out to be an hour or two later than he'd said, the extreme scrutiny with which he'd insist that we examine every sentence, the further rewrites of chapters we'd already written, the illnesses that would stop us in our tracks. When in Cambridge I'd start joining his carers on their cigarette breaks, needing, as I never have, before or since, the psychological crutch and energy infusion that you get from a few drags off an unfiltered Camel. Our deadline would be extended two or three times. But our progress, though slow, was steady, and over time both our page count and our relationship continued to grow, and I never had to have that sort of talk with him again.

9

It was 1985. Peter Guzzardi sat in his cheap rental car in the parking lot of a cheap hotel, waiting for the author who was staying there. He was used to cheap hotels. Like most New York editors, he always traveled discount. Publishing is not a high margin business. This trip had taken Guzzardi to Chicago, where it was hot and muggy. Springtime is the best time in Chicago, but that doesn't mean much. This was May, and the city was already a steam room.

None of that bothered Guzzardi. He was forty and just happy to have made it into publishing. He had a senior editor's position, but he hadn't exactly reached the top. His company, Bantam, was only a couple of years into its metamorphosis from a cheap paperback novel purveyor to a respected publishing powerhouse.

The author Guzzardi was waiting for would be a major contributor to that transformation, but Guzzardi couldn't have known that yet—he hadn't met the guy or seen any of the book he had started to write. When Guzzardi heard that he would be traveling from Cambridge to Chicago to give a talk at Fermilab, a nearby particle accelerator, he decided to fly in to introduce himself and to go over with him the latest draft of the publishing contract. They'd agreed on terms in principle, but it hadn't yet been signed.

All Guzzardi knew about the author was what he'd learned in a *New York Times* profile of him. The article described his passion for physics and for bringing it to a popular audience—and his apparent love of the limelight. All that had made Guzzardi think *This guy should write a book for us.* And now he was. It would fit squarely into Bantam's ambition to turn itself into a respected publisher.

The book was to be called *A Brief History of Time,* and it was to be a history not so much of time as of cosmology, and the attempt to create a theory of quantum gravity that would carry the promise of unifying all of fundamental physics into a single theory. That was heavy stuff, and back in 1985 it wasn't clear that anyone would want to read about it. Back then the author's name was not a draw. Despite that *New York Times* profile, few outside the physics world had heard of Stephen Hawking.

The market for popular science books hadn't yet taken off. Still, every few years there had been a success. In 1977 there had been a breakthrough for *The First Three Minutes,* about the big bang and its aftermath. In 1980 came Carl Sagan's *Cosmos,* but he was already a television celebrity. And in early 1985 Feynman published his first book of anecdotes, *Surely You're Joking, Mr. Feynman,* a surprise hit. So maybe Guzzardi was onto something. But those books were all accessible and extremely well written. No one knew how this book would turn out, and the proposal had been uneven. Some parts were overly technical while others were overly simplistic.

Back at Bantam, opinion on the project was split. What *was* pretty much agreed upon was that they were overpaying. They'd ended up in a bidding war for the book and won it with an offer of $250,000. Cambridge University Press, to which Stephen had first considered selling the rights, had offered a tenth of that amount. Some at Bantam saw in *Brief History* great potential, some not so much, but even if the book were to lose money it would bring Bantam prestige, and so they'd made the offer.

As Guzzardi waited, Stephen was on his way back from the lecture at Fermilab. The topic had been the no-boundary proposal, which had quickly become his pet theory. The stage apparently wasn't wheelchair-accessible, because the *Chicago Tribune* would report that, upon Stephen's entrance, the hundreds of physicists in the audience "suddenly fell silent when they realized that the limp, doll-like object being carried into the auditorium by two men was Hawking." It made for good drama, but to many in that audience, the topic Stephen would talk about was as startling as his physical appearance: here was the person who'd made his name proving that the universe began with a singularity now proclaiming that the singularity, due to quantum effects, is not really there.

That was just Stephen being Stephen—after all, he'd also preached that black holes could not radiate and then proved that they do. "Most scientists, when they do a major piece of work like singularities, hang on to it and resist change," said Michael Turner, a well-known astrophysicist on the staff of both Fermilab and the University of Chicago. "Hawking is more than willing to prove his own work wrong. He is unique."

It was the story of Stephen's intellectual quest to understand our beginnings—with all its ups and downs and flip-flops—that Guzzardi hoped would be the heart of *A Brief History of Time*. He had believed from the start that it would be the personal tale rather than the historical or the technical content that would give the book its appeal. Of course, making a commercial success of the book would work only if he could get Stephen to make the text readable.

Told that Stephen would be arriving shortly, Guzzardi

waited in his car. After a while another cheap rental vehicle pulled up and parked nearby. Guzzardi watched its young driver get out and extract a wheelchair. Then the guy walked to the passenger side, opened the front door, and leaned in. When he straightened up he had in his arms what Guzzardi would later say "looked like a scarecrow." He kept watching as the fellow carried the scarecrow to the wheelchair, lowered it in, and sat it up. He lifted its right hand. Slowly and with precision, he positioned the hand on a knob that controlled the chair. The driver was a twentysomething graduate student who'd made the trip to help care for Stephen and to translate his garbled speech. Stephen was now in position and ready to go.

The wheelchair immediately spun two 360s, then raced off for the hotel entrance. The grad student saw Guzzardi observing all this. "Peter Guzzardi?" he asked. Guzzardi nodded. "That is Stephen Hawking," he said. Guzzardi and the student took off after Stephen. They had to run to keep up, though it wasn't the kind of place you'd think anyone would be in a rush to get to.

Guzzardi followed the student and Stephen up to his tiny room. Stephen was famous among physicists and there had been a few media articles about him, but he wasn't yet a celebrity or well-heeled, so he still stayed in places whose signs advertised "air conditioning" because by the look of them it appeared there might not be. Despite the modesty of the surroundings, Guzzardi was intimidated. He'd anticipated that Stephen would probably be the smartest person he'd ever meet, and that he'd be formidable and imperious.

When they entered the hotel room Stephen muttered a few words to the grad student. To Guzzardi, Stephen sounded like "Darth Vader with a head cold." He didn't understand any of what Stephen had said, and the student didn't translate. Nervous, Guzzardi was anxious to get off on the right foot. He decided to break the ice and start talking himself.

"Hi! So happy you're here!" he began. "Great to meet you! Hope your lecture here was a great success! And I hope you were comfortable in your travels!"

It was just small talk, but Guzzardi was making it with a smile and enthusiasm. Stephen answered, again incomprehensibly. This time the student repeated Stephen's garbled words so Guzzardi could understand them.

"Did you bring the contract?" had been Stephen's reply.

After Stephen and I had our talk, the rest of his Caltech visit had gone well. In the months since, we'd both been generating pages at a decent rate. Now, back in Cambridge, I was sitting at his side day after day, going over what we'd each written since that Caltech meeting.

At one point I asked a question on a physics issue, and Stephen didn't, as was usual, immediately answer. It seemed I had stumped him, and that felt strangely good. But then I realized that, actually, he had fallen asleep. I later found out that others had had a similar experience. In this case, Stephen had been out late in London the night before; the day after such excursions was often a slow one. When he opened his eyes again after a couple of minutes, I asked if he wanted some coffee. He indicated yes.

Dawn, his carer, was sitting on the orange couch with her face buried in a magazine. She'd certainly heard the conversation but she didn't get up, so I stepped to the counter to make the coffee. She was in a grumpy mood because plans were being made for an upcoming foreign trip, and she wasn't on the roster. At least not yet. But the full team hadn't been chosen. Stephen's decisions seemed to never quite be final.

Most of the carers liked going on the foreign trips. They jockeyed to be chosen for them. The destinations were usually fun or exotic, and their expenses were covered by the host institution. So the accommodations were nice, and they got "combat pay" for agreeing to leave their lives behind and come along. Although there was a head carer charged with scheduling the others, both in Cambridge and on trips, Stephen often intervened, and everyone knew it. So they played to Stephen and ignored the head carer. Her job was like that of a cashier in a restaurant where the food was free. Not much point, and she hated it. Sam, though his official role was computer and technology assistant, was also involved in coordinating travel, but he was more detached. He'd watch it all unfold and just roll his eyes.

For the carers, playing up to Stephen wasn't hard to do. They had plenty of face time to argue their case. He couldn't walk away, or even turn away. The only thing he could do was grimace. Stephen could make a grimace feel like a lion's roar, but he was frugal with histrionics. All his carers needed to employ was a little charm and they'd have a good chance of getting what they wanted. It might seem he was easily manipulated, but he knew the score. If he gave them what they wanted, it was because Stephen cared for his carers.

The coffee was ready. Just as I poured it, Dawn suddenly popped up from the couch. "Leonard! You don't need to do that! Allow me," she said, as if she hadn't heard his request. With that she took over, donned her best smile, and brought the coffee to Stephen.

"Out late and paying for it, aren't you, Stephen?" she said as she spooned some of the hot liquid into his mouth. She'd pulled her friendliest voice off the rack to go with the smile, but it didn't fool me. It was like the treatment my kids would give me when they wanted ice cream. Stephen understood that, but still, he basked in the attention. He answered *yes* with his face. He *was* paying for having been out late. He raised his eyebrows and smiled broadly—he clearly thought the late night had been worth it.

"Your problem is that everybody wants to be with you, and you can't say no to a party," Dawn said. I didn't know whether Dawn had a college degree, but if she did, it must have been in flattery.

Several spoonfuls later Judith stepped in. She'd noticed the coffee break and decided to take advantage of it. Unlike many of the carers, she was never obsequious. "Sorry, Leonard," she said. "Just a few things I must squeeze in." My work with Stephen fell down another notch in the queue. She knew I didn't like that. Many times I'd pop into her office to chat—and complain—when others interrupted. But unlike many of the others, Judith always did her homework before interrupting, so she made very efficient use of the time. Anyway, I was impressed with the degree to which Stephen would clear his schedule when I was in town, and I knew I didn't have much to complain about.

Judith had a whole slate of issues to go over. Since Stephen could quickly respond yes or no with his face, she phrased all the issues as if they were part of a game of twenty questions. *Do you want to meet with the fundraiser? Monday? Tuesday? Wednesday? Yes, Wednesday? Does 3:00 p.m. Wednesday work?* The topics she raised covered quite a lot of ground, having to do with everything from the press, research grants, conferences, travel, and invitations to financial dealings. One issue might concern whether he wanted to respond publicly to something provocative that a member of Parliament had said, and the next could be in regard to confirming the time for his daughter, Lucy's Sunday visit.

Judith's last question was whether Stephen was ready to sign some legal document. She had brought her inkpad, and when he responded *yes,* she rolled his thumb on it and then on the contract. Under the thumbprint she wrote "Witnessed by Judith Croasdell" and the date. With that, she retreated back to her office. By then I'd become accustomed to the lack of privacy in Stephen's life, but now I was glimpsing how much he was at the mercy of those who surrounded him. To see how Stephen "signed" documents was to realize how easy it would have been to hijack that thumbprint. Good thing he had, in Judith, someone he could trust.

All told, there had been a twenty-minute delay. But the coffee perked Stephen up, and we got back to work. We'd been discussing *A Brief History of Time,* which had been about how the universe began and how it has evolved since then. Parts of the book were devoted to explaining how we'd discovered what we knew, especially Stephen's contribution, but the science in *Brief History* was the science of the 1980s. *The Grand Design* was meant to go a step deeper. It was based on newer science, on a (then) ten-year-old theory called M-theory and on Stephen's work in the early 2000s. What's more, in *The Grand Design* we intended to address the issues of why the laws of nature are what they are, and why the universe even exists, which would lead us to answer the question *Did the universe need a creator?,* which Stephen had posed in the conclusion of *Brief History.*

When Stephen finally got back to communicating, he typed out, "We have to keep in mind that our book is about whether there is a grand design—a set of laws to control the universe. This involves the God question."

"In *Brief History* you ended by saying that if we physicists could arrive at a unified theory, we would then 'know the mind of God.' You gave the impression that you might believe in a kind of god," I said.

He grimaced. He was saying *no.* Either he didn't think he gave that impression, or he didn't mean to give it.

"Not the biblical God," I said, "but god as the embodiment of the laws of nature."

He started typing. "You can define God that way but that is misleading. It is not what people usually mean by God," he said. "It is redundant because calling it God doesn't add anything."

"We don't need to address the issue directly," I said. "But by saying that there are no exceptions to the laws of physics, we're saying that we don't accept the usual definition of God, in which God intervenes in people's lives."

"I don't want to be labeled either as atheist or deist," he said.

"But you will be," I said.

Just then Patrick came in. He was the carer who had the next shift.

"Hello, Stephen," he said. "How are you today?"

With all the people around Stephen being used to interrupting at will, I wondered how he ever got anything done. On the other hand, Stephen didn't feel obliged to answer. He continued with no acknowledgment to Patrick.

"Our point is that scientists have to believe that the laws of nature always hold," Stephen said. "That's not faith. It is based on experience."

"I just wanted to thank you for putting me on the list for the trip," Patrick said, ignoring the fact that he was being ignored.

Now Dawn jumped in. "How did you get on the list?" she asked. She looked at Stephen. "You said you hadn't decided!"

As Patrick tried to explain why it was vital that he be included, he and Dawn finally had Stephen's attention. Stephen watched as, back and forth, they bickered about who'd be best to take along, and what promises had been made. Stephen could have ended it with a simple word or two, but he just watched. There was often theater among Stephen's carers. He hired a lot of dramatic types. To him, the carers played many roles, and one of them was that of furnishing his own private soap opera. I, on the other hand, was getting frustrated. First the coffee, then Judith, and now this. I gave Dawn and Patrick a wave. They got the message and quieted down.

Stephen went back to typing. He had a lot to say, for him, and took another twenty minutes to get it out. I watched as he typed. Although I was feeling impatient, I didn't try to help with guesses to complete his sentences. He was addressing an important subject, and I wanted him to say this in his own words. Eventually I looked away from his screen, trying to empty my mind and relax. Finally he selected the speak option and his computer voice read what he'd typed.

"A law is not a law if you allow God to intervene," he said. "That leaves two possible roles for God. One is to choose the initial condition of the universe. We got rid of that with the no-boundary proposal. I wrote about that in *A Brief History of Time*. The other possible role is for God to be responsible for choosing the laws and making a universe based on those laws. It shall arise in the last chapter of our book that it is not necessary to invoke God for this either." The argument for that position was based on Stephen's most recent work.

"I want to make my position clear without being violently antireligion like Richard Dawkins," he continued. "I have just been sent his book *The God Delusion*. If you want to see it, get Judith. I agree with most of it, but I don't think it is necessary to be so aggressive."

Stephen was sensitive about not wanting to insult or anger readers who believed in God. Though he didn't talk about it, I thought he must have had the same concern regarding his family. Even so, Jane had written after their divorce that "belief in the higher influence was my continued source of help and strength . . . Was I obliged to let physics, the epitome of rational thought, through its contempt and disdain, destroy the essential motivation of my life?" I can safely say that Stephen had no contempt or disdain for those who believed in a "higher influence," and it was unfortunate that Jane felt that attitude was ingrained in the field of physics. Certainly, destroying "the essential motivation" of your wife's life does not bode well for the marriage.

I found Jane's take on Stephen's attitude toward religion puzzling. I knew that Elaine hadn't felt as Jane had in that regard. Elaine was equally religious, maybe more so. A Protestant, when she'd been asked to accompany Stephen on a trip to meet the pope, she'd agreed only with great reluctance, and said she'd refuse to shake his hand. And when she and Stephen were engaged to be married, she told Stephen, "You're not going to be first in my life because God always is." To that, Stephen replied, "I don't mind coming second to God."

If general relativity and quantum theory could get along, so could Stephen and a deity. After he and Elaine were married, Stephen would often accompany her to church. More than once he was so moved by the service that he cried. At home, Elaine would bow her forehead against his, or hold his hand, and pray at his side. He'd sometimes ask her to pray for Lucy or for a grandchild. Other times she'd pray for his health.

However, I'd recently heard that, after more than a decade of matrimony and another decade of relationship before that, Elaine's marriage to Stephen was now on the rocks. I couldn't help hearing about it, because the carers liked to talk. If they were his soap opera, he was also theirs. But one thing I knew was that it wasn't religion that had come between them.

Some of the carer talk was pretty alarming, and some of it was almost certainly not true. There had even been rumors a couple of years earlier that Elaine was abusing Stephen. Various minor injuries he'd had were supposedly her doing. A cut lip. A black eye. She'd allegedly let Stephen sit too low in the bath, so the water ran into the stoma in his throat. Stephen's world was split into two camps on this. His son Tim and Lucy appeared to believe the stories. So did many of Stephen's carers. Stephen's sister, Mary, and his friends Kip and Robert Donovan said they didn't. More important, Stephen himself vehemently denied that he'd been mistreated, and a police investigation found "no evidence to substantiate any assertion that anyone has perpetrated any criminal acts against Professor Hawking."

Whatever happened or didn't happen, one thing everyone did agree upon was that Elaine and Stephen had always had a stormy relationship. One moment it was *You're crazy, I hate you, and never want to see you again;* the next it was *I love you more than anything and could never live without you.*

I'd met Elaine a few years earlier at the Frankfurt book fair, where Stephen and I had gone in support of *A Briefer History of Time.* But so far, in my dinners at Stephen's house I hadn't seen her. She'd either been away or stayed upstairs while I was there. That was about to change. When we were done for the day Stephen invited me home with him as usual, and he said that this time Elaine would be doing the cooking.

Stephen once called the house he lived in with Elaine "the house that *A Brief History of Time* built." The home had reportedly cost $3.6 million. That's a lot, but if it was bought with *Brief History* money, there would have been plenty left over. Not so much, though, that the cost of Stephen's care couldn't suck it all up. He was worth millions, but he typically had nine or so carers on staff, costing hundreds of thousands of dollars each year. His MacArthur funding had eventually ceased, and book royalties tend to decline over time, so to feed the medical-care beast Stephen was constantly in search of more revenue streams.

Stephen knew that if he lived, the cost of his care would only go up as he aged while his ability to earn money would diminish. It was his vision of that need in his future that had inspired him, in the early 1980s, to start work on *A Brief History of Time*—a couple years before Bantam and Peter Guzzardi entered the picture.

Previously Stephen had written *The Large Scale Structure of Space-time* for the Cambridge University Press. The press's science director, astronomer Simon Mitton, had since the late 1970s been trying to interest Stephen in writing an account of cosmology aimed at the general reader. In 1982, facing mounting medical bills, Stephen finally decided that this would be a good idea. He wrote a draft of one section of the proposed book and showed it to Mitton. Though as an academic press Cambridge University Press published technical books, Mitton wanted something accessible. Stephen's pages read like a textbook, and the rewrite Stephen drafted wasn't much better. Both versions were loaded with equations. That's when Mitton muttered his now famous warning *Every equation in the book will halve the sales*. Did Stephen not realize that the "general public" didn't have advanced math degrees?

Despite his dissatisfaction with Stephen's sample drafts, Mitton was confident that there was an appetite for a popular book on cosmology. He talked to others at the press, and they decided to make Stephen an offer. Stephen made it clear that he wouldn't accept the usual tiny advance. If money was short in what is called trade publishing—publishing for the general audience—it was almost nonexistent in academic publishing.

Still, after a bit of haggling Mitton and Stephen agreed on $25,000. It was by far the largest advance the Cambridge press had ever offered.

Mitton drafted the contract, but he didn't do as Guzzardi later would and personally deliver it for his signature before Stephen could change his mind. Instead, Mitton had it sent to Stephen's office. That proved to be a fatal error. Stephen never signed the contract, and he never again spoke to Mitton about the book.

Life, like quantum physics, is full of uncertainty, and what is squarely aimed at one point sometimes lands somewhere else. Stephen was well along the road to signing on to write *A Brief History of Time* as an academic press book—a book that would have had a high price point, almost no marketing, and been considered a huge success if it sold ten thousand copies. He'd agreed to a contract, received it, and would have signed it, if not for the chain of events started by that *New York Times* portrait Guzzardi had read.

In addition to Guzzardi, at least one other New York publishing figure had read the profile and also concluded, *This guy should write a book:* Al Zuckerman, president of Writers House, a then ten-year-old New York literary agency. Spurred by this idea, Zuckerman contacted Stephen and by chance reached him before he'd signed the Cambridge contract. Zuckerman asked Stephen to hold off; he thought he could get more money. He had Stephen turn his writing sample into a book proposal, sent it to Bantam and other New York publishers, and orchestrated the bidding war that Guzzardi would eventually win.

It wasn't just the selling power of a commercial publisher and a ten-times-greater advance that Zuckerman secured for Stephen. It was also Guzzardi himself, a brilliant editor who could devote much more time to the book than an academic press editor could afford, and who knew how to cater to a much broader audience.

Over the four years they'd work together on the book, Guzzardi would be ruthless, sending Stephen page after page of editorial notes on each draft. Guzzardi wanted the book to sound conversational and personal, and he found Stephen's drafts suffered from the same malady Guzzardi had identified in his proposal. Long stretches of the text were dry and uninteresting, and the tone was uneven, with some passages seemingly written for a twelve-year-old and others for a physics graduate student—or even one of Stephen's peers. To Stephen's great annoyance Guzzardi critiqued the drafts relentlessly. In the end, the book was still a tough read for a popular audience, but if you worked at it you'd be able to understand the gist. Stephen had delivered. When *A Brief History of Time* was published, in 1988, Bantam couldn't keep up with the demand. Worldwide it has now sold well over ten million copies.

Heading home after our day's work, Stephen and I got into his van along with Patrick, who'd do the driving. The van was a custom job. Stephen could have had a fancy sports car for what it cost, but he couldn't have gotten into it. For a vehicle to be accessible to him, it needed major modification. The passenger-side seats in the van had to be removed. A metal ramp had to be built in. And a motor was installed to raise and lower the ramp. The idea was for Stephen's wheelchair to be rolled up and into the van behind where the front passenger seat would have been. The clearances were a little dicey, but if you moved carefully, it worked. Once he was inside, you turned the chair toward the windshield and rolled him forward. There were a series of straps and metal hooks you had to affix to secure the chair in case of bumps or an accident. The chair also had a Velcro strap to hold his head in position. All of this was necessary before we could make the five-minute, mile-and-a-half drive to Stephen's home at 23 Wordsworth Grove.

After pulling into the driveway, Patrick got Stephen out of the car and we made our way up to the door. The house had an understated grandeur that was more indicative of Stephen's celebrity than his Cambridge office. I rang the bell as Patrick returned to the car to fetch the heavy medical bags. Elaine answered. Stephen was then in his midsixties and Elaine in her midfifties.

"Hi, Elaine," I said. "Good to see you again" was about to come out of my mouth, but she preempted my attempt at a cheerful greeting.

Ignoring me, she glared at Stephen. "Who is he?" she said. She sounded angry.

Then, to me, "Who are you!"

"I'm Leonard." I felt apologetic. "We met in—"

"*Who* are you!"

"I'm writing a book with Stephen—"

"You're an editor?"

"No, I'm his coauthor. I—"

She ignored me and turned to Stephen. "You brought *him* for dinner?" She said it with the indignant tone you might use if you were whispering it confidentially in his ear, but she was practically yelling.

"It might have been nice to let me know," she continued. "You never do, do you! Because you're Stephen Hawking, and you don't need to! Well, there's not enough food!"

With that I started to withdraw, but I could tell from Stephen's eyes that he wanted me to stay. He projected no sign of shame or apology. He even flashed a quick smile, as if Elaine had said, *Hi honey! So nice that you brought a friend for dinner!*

I told Elaine I wasn't really that hungry but that I'd be glad to sit with them and keep Stephen company. She grew calmer. This wasn't my fault, she said. She ignored Stephen and invited me inside. Just then Patrick showed up with the medical bags and wheeled Stephen off to the restroom.

"I'm sorry," Elaine said after they'd left the room. "It's just that I've been his slave for twenty years, and it's enough."

Elaine walked me into the dining area outside the kitchen. I got the wine from the closet. Patrick rolled Stephen in to join us, and Elaine served the first course. No one except Patrick was saying much. He looked confused. He could sense that something tense had transpired, but since he hadn't been present for it, he didn't know what. He kept feeding Stephen and tried to go on as if this were a normal dinner. He made some small talk.

After we'd been eating for a few minutes Elaine grabbed her plate and stood up. "I can't handle this," she said. With that she walked off, plate in hand, and headed up the stairs. I was mystified.

It wasn't the first time she'd had that effect on me. When I first met Elaine in Frankfurt, I'd wanted to take her picture. Although this was out in public, she reacted as if I were a Peeping Tom. "No!" she'd shrieked. At first I'd thought she was angry, but it soon became clear she'd been mortified. "I'm a nothing, a nobody!" she went on. I apologized profusely, and then she softened, as if she realized she'd overreacted and wanted to explain herself. "I'm sorry," she said. "I just don't want my picture taken. I'm nothing. I'm invisible—like the air." I wondered if she was about to cry.

That incident, too, had left me puzzled. I'd stepped on a mine, but why was it there? Did she consider a photo an invasion of privacy? I knew from her reputation that she wasn't shy or an introvert, so why make a big deal of it? Had it been resentment? Was she saying *How dare anyone pay attention to me now—too little, too late?*

Years later, when I spoke to Elaine after Stephen's death, she seemed to have mellowed and was able to give me some insight into the scenes I had witnessed. "Stephen was like an actor," she said. "He needed to be the center of attention, the center of the universe. He loved it. It gave him energy. He loved people. He had a very tough life but he was an incredibly brave man. He never, ever complained, ever, but he needed to be the center of attention. And, yes, I probably resented that. Not all the time, but when I was tired or one of the carers was flirting with him, or whatever it was. But it would be temporary. The resentment would pass. Deep down, he was my only love," she said. I believed her.

There are many words one could use to describe Stephen. Courageous. Stubborn. Skeptical. Visual. Passionate. Playful. Determined. Brilliant. Fun-loving. But when you got to know him intimately you realized that what he was most of all was vulnerable. His sweat glands could torture him. His stoma could suffocate him. His friends or wife could betray him. He was at the mercy of everyone and everything, and as our relationship deepened I grew to admire the grace with which he accepted this. But there was also a flip side. To be associated with Stephen made you vulnerable, too. His bodily needs could interrupt at any time. Plans could change in an instant. Chaos was normal. Time was impossible to manage. Communication was spotty and could take forever. The first priority was always physics, and undone tasks always quickly piled up. Thank-yous were few and far between. And to any new bride, there was the baggage of his first marriage.

To be married to Stephen you surely had to give up a part of yourself. It's not that Stephen had a cold heart, but that his warm soul was captive. I wondered, *How good could it feel to hold him, when he couldn't hold you back?* With the difficulty of that connection and the sacrifices you'd be called upon to make due to his needs as well as his fame, it'd be easy to be subsumed, to lose direction, to feel small.

I felt much affection for Stephen and found our work immensely rewarding. But I would not have wanted to be his roommate, nurse, or wife. I imagined that living with him could drive a person crazy. I imagined that that's what was happening to Elaine, at least on and off. Maybe to Jane, too. I couldn't judge. I met Jane only a couple of times.

Elaine's abrupt departure from the dinner table left us all feeling awkward. I didn't think anything we'd said had triggered it. Her earlier upset must have simply kept bubbling until it finally boiled over. I didn't know how I was supposed to react, and Patrick certainly couldn't have known. "Nice salad!" he said finally, and he continued feeding Stephen.

It had been quite an evening, and when we'd finished serving ourselves the dinner Elaine had prepared, I was happy to be leaving. I had in mind a few beers at that late-night pub I'd found, and I was anxious to depart so that I could make the half-hour walk and still get there before the doors—but not the pub—closed at eleven. I said my goodbyes, but just as I was leaving, Stephen seemed to have something to say. I waited as he typed.

"Our advance is not enough money," his voice said.

I shrugged. "I thought it was a little low myself, but I figured you and Al were the ones in charge of that part." Al was Al Zuckerman. Thirty years after *Brief History,* he was still Stephen's agent. He started typing again. It seemed odd that he was bringing this up now.

"We should have gotten double," Stephen said.

I laughed. "I think so, too! Too bad," I said. "I'd wondered why you hadn't demanded more."

Stephen grimaced. That confused me, since I had just agreed with him.

And then he said, "I want you to tell Al we want our advance doubled."

This threw me off. How could we tell Al to ask for more? We'd all agreed to the advance. We had a contract. We were far into writing the book. Stephen's suggestion seemed as absurd as the idea that a black hole radiates.

"Uh . . . I don't see how we can do that," I said.

Usually pretty confident, I now felt sheepish. It felt wrong to go back on our word and our contract. But I also felt uncomfortable telling Stephen I didn't want to. He grimaced again.

"Al will be very upset," I said. "No one does things like that."

"If Al doesn't like it, bugger Al," Stephen said. I knew "bugger" was a mild British swear word. I'd met people in pubs who'd wanted the prime minister buggered. But I didn't really know what it meant, and it was odd hearing a computer voice curse. It was also odd to hear Stephen talk that way about the person to whom he owed his writing career. Still, the gist was clear.

I thought about it for a moment. He seemed pretty adamant. Finally, I said, "Okay. I was going to stop in New York on my way home. I'll talk to Al and let you know."

Patrick, who had been observing all this, seemed to find it amusing.

"Welcome to Hawking world," he said. And then, to Stephen, "Can I get my pay doubled, too?"

10

When *A Brief History of Time* came out, on April Fool's Day, 1988, Stephen, then forty-six, was already considered by his peers to be one of the greatest theoretical physicists of his generation. Had he instead been one of the greatest basketball players, or singers, or CEOs, he'd have earned enough to be set for life. But for Stephen, before April Fool's Day, 1988, supporting himself had been a ceaseless struggle. It wasn't about paying the rent, it was about staying alive. Being *kept* alive. A candle, once lit, is supposed to burn steadily till the end. But Stephen's candle needed constant tending. Hour after hour, day after day, year after year, there was always the chance that it would be quenched by a stray puff of air.

Though an accurate headline describing *A Brief History of Time* would have been something modest, like *Leading Physicist Explains His Theories,* that's not the way the book and its author were described in the media. In the media, Stephen Hawking, the man who couldn't move, was called *Master of the Universe.* Of Stephen Hawking the atheist, it was declared that *Courageous Physicist Knows the Mind of God.* The inflated headlines were merely the media marketing its own articles. It's a business in which an academic paper reporting that the sun will explode into red giant status in five billion years might be announced with the headline *Scientists Say World Coming to an End*. But the hype didn't just market the articles; it marketed the book, and it marketed Stephen.

In the eyes of the public Stephen quickly became not one of the greatest physicists of his generation, but one of the greatest minds since Plato. Some of Stephen's colleagues chuckled at the hype and felt happy for him. Others groused. One said, in a 1988 London *Sunday Times* article, that a list of the twelve best physicists of the twentieth century would not include Stephen. Stephen would have agreed with that assessment. He knew that, back in his early days at Caltech, he would have come in at just number three among the physicists on the fourth floor of the Lauritsen Laboratory—behind Murray Gell-Mann and Richard Feynman. Still, Stephen was happy to have become a spokesperson for his field. What was more important was that probably more than any other physicist alive, he needed the money that comes with fame. Psychologists argue over whether money brings happiness. For Stephen, money brought life.

Stephen's fame didn't go to his head. He'd always had a certain arrogance—as do most people who are that smart and accomplished—but he also appreciated that, smart as he was, nature is smarter, as all theorists know from experience. The book *did* change Stephen's physics life, however, because his newfound celebrity sucked up so much of his time. Given the intense media activity and the many invitations he'd accept after the publication of *Brief History*—not to mention his breakup with Jane, his subsequent marriage to Elaine, and the business of moving into a new house with her—the 1990s would be one of his least productive periods as a physicist.

Stephen's most famous contribution to the field in the decade after the publication of *Brief History* wasn't a feat of physics, but rather of physics marketing. It was caused by the stir arising from a 1997 bet that he and Kip Thorne made with Kip's Caltech colleague John Preskill. The bet concerned an issue Stephen had first raised in 1975.

From the point of view of physics, there is information encoded in all matter. A helium atom, for example, carries the information that it is not hydrogen or some other element. Stephen's bet concerned the fate of that information after a bit of matter becomes part of a black hole that later vanishes through the process of Hawking radiation. The issue is often called the black hole information paradox. Due to Stephen's reputation, his bet about it generated headlines around the world and added to a revival of interest in the issue among physicists themselves.

Physics is about predicting the future. Not the future of human society, or the stock market—those are too complicated and must be left for other disciplines to tackle. We physicists focus on matter and energy in its simplest forms. Particles. Light. Materials. Fluids. We create theories of those things and derive the laws they obey in the service of being able to understand how systems of matter and energy interact and evolve over time.

Given that the central goal of physics is prediction, it's no surprise that a basic requirement of theories in physics is that they tell you, given a system's current state, how to calculate its future state. That's where information enters the game—by "state," physicists mean the relevant data about an object, and data is information.

We've seen that, in quantum theory, the information relevant to a system is encoded in its wave function. That wave function changes over time in a manner that reflects how the state of the system will evolve, and if you know the wave function at any one time, the rules of the theory tell you how to calculate what the wave function will be at any other time. So if you know the present wave function for an atom and want to know the probability that it will have certain properties a minute later, you can extract that from the wave function.

Just as important, you can run the calculation backward—from the wave function at a later time, you can reconstruct it at an earlier time. As far as the wave function is concerned, both the past and future are knowable. Physicists call that property unitary evolution, or, more simply, unitarity. It is one of the most fundamental principles of both the mathematics and the physics of quantum theory.

Sand, stirred into water, will not make it salty. Salt, on a beach, would dissolve into the ocean surf. The stuff of nature can be transformed, but each material, each molecule, atom, particle, has its own identity and characteristics, and its own reaction to being dropped into water, or burned, or smashed. In principle, even the smoke from two burning books will differ in a manner that reflects their initial identity. That's a consequence of unitarity—it means that by analyzing the result of a process, you can (in principle) infer how the system started out: if water is salty, you know that salt, not sand, was stirred into your glass.

That's where black holes seem to differ from any other object in the universe. If grains of salt and sand were to be tossed into a black hole, they would both increase its mass just a bit, but change it in no other way. The characteristics that distinguished one substance from the other therefore no longer exist. And, since both substances would have the identical effect, from the outside there is no way you can later determine what had fallen in. That's a problem for unitarity, because it means that if a black hole swallows matter, you can no longer use information about the current state of the system to reconstruct its past. The past is no longer discoverable. It has been erased.

But does a black hole really swallow up matter? Consider a thought experiment. Suppose that Kip and Stephen are each in spaceships, exploring space at some distance from a black hole.

Kip decides he wants to see what the black hole's inside looks like, so he lets his ship plunge toward it, and he takes note of what he sees after he passes the black hole's horizon. Sadly, he'll have to keep whatever he discovers to himself, because once past the horizon, neither he nor any messages he broadcasts can ever get out. That's a scenario often talked about in popular descriptions of black hole physics. But Kip's perspective is not relevant here. What's relevant to the issue of information loss is Stephen's perspective, the perspective of someone on the outside.

From Stephen's perspective, Kip will never fall into the black hole. In fact, those who keep their distance from a black hole will never observe any object falling in. That's because as perceived by those at a distance, time near the black hole slows down. Distant observers will see clocks tick more and more slowly as those clocks approach the black hole. Similarly, they'll see objects approaching the black hole move ever more slowly—until they are moving so slowly they will seem to not be moving at all.* So although an observer like Kip can, from his own point of view, fall into a black hole and look around inside it, from the point of view of Stephen, who stays a distance away, all objects, including Kip, will seem to slow down and stop just outside the black hole. It will seem as if they become "stuck" to its surface.

That the two observers experience contradictory versions of events is strange. But it is not a problem for physics, because those who fall in and those who remain on the outside cannot communicate with each other—it is as if they exist in two separate, parallel universes.

---

* General relativity also tells us that the slowing down of time just outside the black hole means that light waves emanating from the objects stuck outside will oscillate ever more slowly. Their frequency will eventually diminish to such an extent that whatever our current state of technology, those objects will be undetectable. In some respects, that makes the issue of whether they have fallen in or remain stuck just outside the black hole horizon a moot point.

What's important with regard to the principle of unitarity is that, as far as we outside observers are concerned, objects never complete the process of falling in. They don't really get swallowed up, so the information they carry is not lost. The principle of unitarity is safe.

That's where Hawking radiation comes in. According to Stephen's calculations, a black hole will radiate energy, and that radiation will be the generic glow given off by any hot body. It will contain no information. What's more, Stephen predicted, as the black hole shrinks, the process will accelerate until, in the end, the black hole disappears in a powerful explosion that leaves no trace. At that point the information has been lost. The principle of unitarity has been violated. The mathematics of quantum theory says that can't happen, but Stephen's black hole theory says it does. That's the black hole information paradox. According to Stephen's theory, there's a point at which the quantum mechanical prescription for tracing the evolution of systems must break down.

Strangely, for a couple of decades the fact that Hawking radiation seemed to violate a basic tenet of quantum theory didn't attract much attention. Then, in the 1990s, interest grew, and it accelerated after Argentinian American physicist Juan Maldacena made a theoretical breakthrough, and Stephen made his famous bet. In the bet, Stephen and Kip took the position that the information is truly lost and that quantum theory would have to be revised in a yet unknown manner to account for the loss. John Preskill bet that, on the contrary, Stephen's calculations had been wrong. He believed that one or another of the mathematical approximations that Stephen made in order to derive the predictions of his theory had had the effect of making it *seem* that the information was lost when it actually wasn't.

In Preskill's opinion, if one could solve the problem exactly—through that theory of quantum gravity that everyone hopes we'll someday have—or if one could find a better way of approximating the solution, then it would be found that the information in question emerges in some way.

Some who shared Preskill's view thought Stephen was wrong about the characteristics of Hawking radiation. Maybe it isn't really generic heat radiation, as Stephen had concluded, but instead a type of radiation that somehow encodes the information. After all, since from the point of view of the outside observer, objects never fall all the way in but, rather, remain stuck "just outside," one might ask, What happens to that shell of matter when the black hole disappears? No one knows. Could it be shown that the evaporation process restores the information held there? No one knows that, either. Another popular theory was that the radiating black hole would not completely vanish, as Stephen had concluded, but would leave a remnant that would contain the information.

As others pondered such ideas, so did Stephen. Then, in 2004—seven years after making the bet—Stephen arranged to issue a major announcement concerning his latest thoughts about the bet. Having settled the issue to his own satisfaction, once again he was taking the physics community by surprise.

As I'd promised Stephen, on my way back from Cambridge I went to see Al Zuckerman at Writers House. The agency had grown in the twenty years since Al "discovered" Stephen and now had two adjacent brick buildings on West Twenty-sixth Street, just off Broadway. Converted residential walk-ups, they were old four-story buildings with few windows. Over time walls had been knocked out, other walls had been added, and the structures had gradually been shaped into one building that housed nearly two dozen literary agents. It had a lot of character. It was also cramped.

Al was as much a fixture as was his building. Now in his seventies, he was a grand old man in the book business, and he dressed like it. He even had bushy eyebrows that fit the look. My agent, Susan Ginsburg, also a member of the Writers House staff, was at the meeting, too.

Susan hadn't told Al what we were meeting about. Maybe if she had, he wouldn't have let us into his office. But there we were. We started out with the usual small talk. As my lips flapped with chitchat about the lousy Cambridge weather, I girded myself to deliver the Cambridge message. Stephen's message. It would be much simpler to get across than Stephen's announcement of Hawking radiation, but I expected it to go over just as poorly. The idea that *black holes don't radiate* had been a mantra of physicists, and Stephen tore it down. *You don't go back on your word and ask to double your agreed-upon advance* sounded to me like a basic principle of agenting, and Stephen didn't recognize that one, either.

"What? I can't tell Bantam that," Al said when I finally explained why I'd come to see him. Then, sure enough, he continued with, "It's just not done. We have a contract with them. You *agreed* to the contract. We gave our *word*."

"I know," I said. I said it rather sheepishly.

It was a strange situation. I knew that if anyone could pull off what Stephen was asking, it was Al, but if there was anyone who would find the idea abhorrent, that was Al, too. I felt bad about the whole thing, but I'd resigned myself to it. I'd resigned myself to demanding more money. It felt strange to *resign* yourself to getting more money.

"Why is he suddenly asking for this?"

"I don't know," I said. "But that's what Stephen wants."

"Bantam will be angry if I ask for more! They'll go crazy," Al predicted.

"Yes, Al, but they'll agree to it, don't you think?" said Susan.

"They won't agree because I won't *ask* them to agree. I just can't do it," Al insisted.

We all sat there for a moment, looking at each other like three TV quiz show contestants, none of whom knows the capital of Botswana. Susan finally broke the silence.

"Why don't we all sleep on it, and we can have a phone call and revisit the subject tomorrow or the next day," Susan said.

"I don't need to sleep on it," Al said. "Just tell Stephen I said no."

"Okay," I said. "But Al, there's something you should know. Between you and me, I think Stephen seems a little dissatisfied."

"What do you mean?" Al asked.

"He made a comment a while ago," I said. I didn't relish bringing it up, but it was true. "He was worried that you've become less aggressive."

"Less aggressive? Aggression doesn't have anything to do with it. A bigger advance won't make any difference. You'll get the same amount of money either way. If not now, in the advance, then later, in royalties. The book is great. It'll sell a million copies. Just tell him I said it's a bad idea, because the advance doesn't matter."

"I can tell him that," I said. "Or you can email him and say it."

"All right, I will," Al retorted.

I considered leaving it at that. But I decided I owed Al full disclosure.

"I should mention that Stephen also said something else," I added. "He said that if Al won't do it, then 'bugger Al.'"

"I was shocked when Len told me," said Susan. "I'm sure

Stephen didn't mean anything by it. It's just one of those British expressions."

"He really said 'bugger Al'?" Al said. He sounded incredulous.

"Sorry," I continued. "It's awkward to pass along. But that's what he said."

I felt bad about sharing this with him. I thought about how hurtful the comment must be, coming from Stephen, someone for whom Al had done so much. There was a long silence. I wished I was somewhere else, in some more comfortable environment, like in the dentist's office having a tooth drilled. Al turned his gaze toward Susan, and then toward me. His bushy eyebrows went up and down. Then he shrugged. "I've been called worse," he said. He smiled and asked how my kids were doing.

That was pretty much the end of our meeting. After hearing the "bugger Al" remark, Al decided to talk to Bantam, to "feel them out." They must have felt pretty good, because the next thing I knew, our advance had been doubled.

When I next talked to Stephen about it, he didn't react much. He seemed to have assumed all along that Bantam would agree and hadn't given the matter any further thought. He seemed to have appreciated, which I didn't at the time, that Stephen Hawking was the man who'd made Bantam Bantam. But Al had been right, too. *The Grand Design* sold well enough that in the long run the doubling of the advance didn't make any difference.

I was with Stephen the day after Elaine and Stephen were in court and she accepted Stephen's offer of a divorce settlement. He had just turned sixty-five. They had filed for divorce the previous October, in 2006, but Elaine still hadn't moved out. That didn't mean they saw a lot of each other—they had pretty much separate domains, hers upstairs, his on the ground floor. Stephen hadn't insisted that Elaine move out because "he's too nice," Judith said. I thought it was more than that. I thought it was because he'd miss her too much. Just because you can't live with someone doesn't mean you can live without her.

Stephen spent much of that day with tears in his eyes. Strangely, we got a lot of work done. Some drown their sorrows with smoke and drink; Stephen did it with physics. Judith and Joan both told me later that having me there helped keep his spirits up, that he found our discussions "inspiring." It was nice to hear, though I took it with a grain of salt. It wasn't the kind of thing he tended to say.

Whether or not Stephen was inspired by our work, on that day at least, I was. We were talking about the "fine-tuning of the universe." I'd just finished reading a technical book on the subject that I'd found revelatory. In it various theorists analyzed models of how the universe would have evolved if the laws of physics were slightly altered in various ways. How much could you change the laws and still have the universe produce life? According to their calculations, there wasn't much wiggle room.

Stephen and I had talked about the fine-tuning issue before, but until I read that book I hadn't realized just how fine the tuning had to be for a world like ours to exist. Apparently a cosmos with stars, planets, carbon atoms, and the other things necessary for life would not have been possible unless the laws were almost exactly what they are. Change the strength of the strong nuclear force by half a percent, the electric force by 4 percent, or the mass of the proton by one part in five hundred, for example, and we wouldn't exist. I pondered how Stephen's physical universe had been turned upside down by his diagnosis years before, yet he had found a way to survive. And his psychological universe had been turned upside down by the divorce from Elaine, yet I felt sure he'd survive that, too. Life in the cosmos, however, was apparently not that resilient.

In between Stephen's waves of grief, we spoke of the existential implications of the fine-tuning research. There seemed to be only two ways to understand the presence of such a delicate balance of particles, forces, and laws. One is by invoking God. In that case you believe that the universe is fine-tuned according to his grand design. The other explanation employs the multiverse concept. In that case you accept that there exist multiple universes, each with different laws. The fine-tuning, then, is not a mystery because in some universes—those very similar to our own—life is possible, while in others it isn't, and given that we do exist, we have obviously been situated in one of those universes where our existence is possible.

It's like finding a community of fish living in a small lake in the midst of a vast desert. Though the hospitable lake is the only place the fish can survive, it's no "lucky miracle" that they find themselves there, for no fish would ever evolve in the hot, dry sand. That's the point of view we present in *The Grand Design*.

It is important to note that Stephen hadn't come to believe in the multiverse as a way of avoiding the need for a god. Rather, his research led him toward the multiverse, independent of the fine-tuning issue. Nevertheless, the implications with regard to fine-tuning intrigued him. Stephen passionately opposed the notion that if science could not yet explain some phenomenon, it must be because that phenomenon is beyond the reach of science. That's one reason he was so excited by his research on the origin of the universe—it spoke to one of the only realms that science hadn't yet addressed. By addressing it, Stephen felt that his research was reinforcing the validity of science itself, and he took pride in that. The fine-tuning issue was part of this.

Though we were both losing ourselves in our discussion, the

undercurrent of sadness in Stephen was hard for me to ignore. I knew he thought that he could never love another the way he loved Elaine, just as I knew that she felt that way about him. I knew he was probably afraid of being alone. I also knew that he'd been ambivalent about the split. He'd finally come to a decision only after having a long talk with Kip and Robert Donovan. Only with their advice and encouragement did he decide that his troubles warranted ending the marriage. In the final analysis, it wasn't clear whether it was he or Elaine who actually initiated the proceedings, but either way, once he'd decided they should split, as Robert told me, Stephen had his ways of orchestrating it. Still, it seemed that Stephen couldn't think of Elaine without his eyes welling up.

After work Stephen asked me as usual to have dinner with him, this time in the Fellows Dining Room at the ancient Gonville & Caius complex in old Cambridge, where I stayed. The buildings, dating back as far as 1353, were built around two main courtyards. My room overlooked one of them. The structures had a lot of character. They also had the quality of heat, plumbing, and electricity you'd expect in a fourteenth-century stone building.

If you knew which building the Fellows Dining Room was in, getting to it was easy. You just went up a set of old wooden stairs labeled FELLOWS STAIRCASE. That set it apart from another staircase across the way, for those less distinguished, and leading to less distinguished rooms—such as where the students ate. For the handicapped, though, there was only one way up. It was a slow, creaky elevator near the base of the Fellows stairs. Its gorgeous wood paneling seemed to have gained character over the ages. But elevator motors don't gain character with age, and the few times I rode up in it, I found it pretty scary. Stephen's wheelchair was a tight fit, so we had to roll him in, push the button, and let him ride up alone. Then we'd walk up and pull him out on the upper floor. This seemed a bit dicey the first time I witnessed it—what if the elevator got stuck?—but Stephen was comfortable with it. It was a routine he'd repeated hundreds of times.

We started the night in an ornate old room, sipping sherry. We'd finish it in an ornate old room sipping port. These were rooms Stephen had been drinking in ever since his graduate student days, though back then he'd have needed a Fellow's invitation. The walls of both rooms were lined with portraits. I studied one. It depicted William Branthwaite, who'd been made master of Caius in 1607. In Branthwaite's time the college had focused on the study of medicine. Ironically, they'd had a rule barring admission to anyone gravely ill. In addition, the college banned invalids or anyone "deformed, dumb, lame, maimed, or a Welshman." Stephen was all those things, except Welsh. He was lucky he hadn't lived in that era. As for me, I was just happy they didn't ban the inebriated, because on my empty stomach, the sherry was going to my head.

In between the before and after drinks we had dinner in yet another ornate room. That was the Fellows Dining Room. It had odd dimensions—very long and somewhat narrow. Its high ceilings were crisscrossed with cream-colored beams on which were painted intricate multicolored patterns. On the outdoor side, windows went all the way up, with Corinthian columns in between. On the interior wall, high up, there was a frieze of bas-relief scenes of warriors engaged in struggle. Greeks fighting Amazons? I wasn't sure.

A walnut dining table ran almost the length of the room. It sat sixty-four. Only ten of us were dining—a respectable group for one of those round tables at a Chinese restaurant, but in this grand room, having ten seats outfitted with place settings and fifty-four empty ones made me feel as if we were guests in a ghost town. I looked at Stephen, who seemed at home. I suppose one man's ghosts are another man's tradition.

The china was elegant. The food was bland and overcooked. Beef tenderloin, carrots, green beans, potatoes, all done in the old, traditional manner that gave old, traditional English food its reputation. The same stuff Branthwaite ate, for all I knew. The service was a throwback, too. Attentive to a fault. If you took a sip from your crystal water glass, a server would be refilling it before you'd finished swallowing. There were three of them for the ten of us, but only two of the servers served.

The third, a thin middle-aged man called the butler, stood at frozen attention, moving nothing but his hands, which he used to point and gesture. He directed the other two, who were women, as if they were his puppets. He did it with a flourish but in complete silence. Once, at one of our family dinners, I asked my young son Alexei to pass a dinner roll. On a whim he tossed it across the table, and I caught it. I didn't mind that. It was fun. This wasn't that type of dinner.

Tonight there was conviviality despite the formality. Once, when I was precipitously losing weight, I had a doctor who said, *Calories are life*—you have to eat to stay alive. For Stephen meals were always more than that. He didn't mind the dryness of the meat. It helped that his carer was mixing it with plenty of gravy as she spooned it to him, but he wasn't just feeding on meat, he was feeding on the company.

I sat with Stephen next to me. On my other side was a chatty fellow who'd once been the British ambassador to Poland. It was a job that had apparently required much vodka drinking. They poured it for you whether you wanted it or not, and you were expected to drink it. He was more sober now, and had the title

Master of Caius College. I was impressed that almost exactly four hundred years after William Branthwaite's time, this man sitting next to me had the same job. The ex-ambassador and current master of the college was a chain talker. He spoke at length about the art of keeping up appearances at official Polish gatherings while secretly dumping your vodka into a potted plant.

The topic seemed to interest Stephen. I remembered that as a student, Stephen had told his friend Robert that the key to getting through fancy but boozy Cambridge dinners without feeling sick the next day was to drink a lot while eating and forgo the port or cognac served after. In Cambridge you can refuse because there is ceremony in the pouring, and it begins with the question *Would you like some?* In Warsaw, they apparently skipped the ceremony. Stephen seemed to have some opinions on all this. He seemed to want to chime in at a couple of points but didn't. He was too busy chewing the tough meat.

Though I figured our vodka-dodging conversation wouldn't meet Cambridge's traditional scholarly standards, to me it was a thrill to know that my path had crossed with that of the Caius master, a William Branthwaite figure of the early twenty-first century—a man who, by virtue of his position, might be thought of by diners sitting here four hundred years hence.

If I'd felt part of Cambridge history simply by being here, I imagined Stephen relished the fact that, through his discoveries, he actually contributed to it. Here at Caius, where it felt to me as if I were on another planet, Stephen was comfortable, content, and happy. It was a place he'd lived and worked in since his graduate school days, since before he was wheelchair-bound. He'd always loved Cambridge, and his fellow Cambridge physicists loved him. He could have left the university for a much better paid faculty position at Caltech or at any other top university, but here he was at home. On our way to dinner, he had seemed still shaken and pale. Here, in a place full of portraits and tradition, I saw him start to relax.

When we got to the port-and-cheese room Stephen had some of the port, but just a few sips. Although he'd been pretty silent at dinner, here he talked a little. At one point he typed something out on his computer, then smiled as he played it for me in his computer voice. His message was "I'm never getting married again."

"Divorce is always the other person's fault," I said. He smiled again.

When Stephen fell in love with Elaine, he knew he had found in her a reflection of his own joy for life. Whatever he did, wherever he went, he knew she was thrilled to be with him, and vice versa. Back then they were burning with feelings, and though trapped in a useless body, his spirit could soar through their love for each other.

Now he faced the future alone. If he were lucky enough to live long and grow old, he'd be growing old without her. But sitting there at that grand table, in that place of tradition and history, Stephen took comfort. Like a religious individual who takes solace in knowing that one's fate lies in the hands of God, Stephen had seemingly always felt soothed in knowing his place in the grand design, in knowing how humankind fit into the plan of nature and the universe as a whole. Now I saw that he was likewise consoled by taking his place in the long and elegant tradition of his university. It gave him perspective. It seemed to help him accept that he'd have to let Elaine go. He knew their life together had passed, just as he knew that in not so many years his own life would pass, that he'd join the other bygone scholars, living on through his immortal ideas and his portrait on the wall.

In 2004, when Stephen decided that he'd resolved the issue of his bet on the black hole information paradox, he chose to present his ideas at the 17th International Conference on General Relativity and Gravitation. Though it had taken him seven years to reach his conclusion, he wasn't going to the conference to claim victory. He was going there to concede the bet. Once again, he'd decided he was wrong on an important matter and was reversing his position.

The conference took place in the grand concert hall of the Royal Dublin Society. It was a gathering much like the 1962 Warsaw meeting that Feynman had attended—and ridiculed. Now, however, some forty-two years later, general relativity and gravity conferences did not attract "a host of dopes," but the best and the brightest. A big part of that was due to Stephen's work in the 1960s and '70s.

Though Stephen had helped build the field, at this conference the crowd was suffused with, as one blogger wrote, "skeptical curiosity" about what Stephen would have to say. Thanks to Stephen's accomplishments, this wouldn't be the piranha tank he'd encountered when he was young and announced his discovery of Hawking radiation. Still, in one attendee's words, none of the physicists in the audience "seemed to believe that Hawking could suddenly shed new light on a problem that has been attacked from many angles for several decades. One reason is that Hawking's best work was done almost thirty years ago."

That was not the attitude of the media. Press passes to the conference were so popular that those who may have been courted to attend past conferences were now discouraged from attending, or were required to apply for one of a limited number of passes. Moreover, the conference organizers spent nearly ten thousand dollars hiring a firm to keep the party crashers out.

Stephen's research was always very complicated and highly technical, even for theoretical physics. Todd Brun, a collaborator of mine who'd gotten his Ph.D. working with Murray Gell-Mann at Caltech, told me he'd taken an advanced course on gravitation from John Preskill while there. The first two quarters of the course were on general relativity. In those quarters, students were able to achieve a basic understanding of that vast subject. The entire third quarter, on the other hand, was devoted to a single topic: Stephen's calculation of Hawking radiation. To me, the fact that Stephen—who couldn't even scribble a single equation—was able to derive a theory that consumed a whole quarter of a graduate school course is as astonishing as Hawking radiation itself.

Stephen's approach to black hole information loss proved to be as complex and creative as his original work on black hole radiation had been. Physicists often study elementary particles by analyzing the results of what are called scattering experiments. In those, one fires two particles, or beams of particles, at each other. When they collide, very complex interactions take place, too complicated to follow in detail. Fortunately, one can test elementary particle theories by merely keeping track of what you send into the region of collision and then studying what you see coming out of it after the turbulent collisions are over. Such analyses are the bread and butter of elementary particle physicists. Stephen adapted that approach to use in studying black holes—just as in his Ph.D. work he'd commandeered Penrose's big bang techniques for the same purpose.

To attack the information loss issue, Stephen imagined sending a slew of particles toward each other in a particular manner, so that when they met they'd have a sufficient concentration of matter and energy to form a black hole. Then he looked at what the theory said would emerge after all the particle interactions had occurred. "One sends in particles and radiation from infinity [i.e., very far away], and measures what comes back out to infinity," he said in his Dublin talk. "One never probes the strong field region in the middle [where the complex interactions take place]."

Though the concept is simple, the analysis is complex. To accomplish it, Stephen used Feynman's method, the sum over histories. Remember, Feynman's method requires you to add up contributions from each of the infinite ways that the process leading up to your measurement might occur—the possible "histories" of the system of particles you are studying. In tracing the possible evolutions of the collision processes he was considering, Stephen said that although the vast majority of the possible histories included the formation of a black hole, in a few of the histories no black hole formed. Stephen said that this was his big epiphany: "I shall show that this possibility allows information to be preserved," he said.

In those histories in which a black hole doesn't form, there is obviously no black hole information loss, and much of his talk was devoted to arguing that when you add up the Feynman sum over all histories, the inclusion of that subset of histories leads to the information being recoverable—the information sneaks back out via the histories in which the black hole does not form. But the easy logic belies some rather nightmarish mathematics—and the calculation that led Stephen to his conclusion was a bit cryptic. For one thing, it depended on several questionable approximations—huge simplifications—that Stephen had to make in order to be able to carry out the math. He presented them in his talk and promised that a paper with all the details would appear later.

After describing his ideas, Stephen conceded the bet. He announced that he'd been wrong, that there is *no* information loss, and that unitarity, and hence quantum theory, is valid. He presented John Preskill with his winnings—an encyclopedia "from which information can be recovered at will."

After the talk Kip, who'd bet along with Stephen that the information *was* lost, refused to follow Stephen's lead and concede. "This looks on the face of it to be a lovely argument," Kip said. "But I haven't seen the details."

John Preskill accepted Stephen's concession and the encyclopedia, but he didn't buy Stephen's argument either. "I'll be honest," he said. "I didn't understand the talk." To be convinced, he said, he too needed to see more details.

Their reactions were typical of those in the crowd. Both those with Stephen and those against him on this issue were waiting for the details. The old Stephen might have been able to provide those details, but the new Stephen, the one who'd decided he didn't have enough time left in life to be rigorous, didn't actually have them. He had formulated his ideas and then assigned a graduate student to do the grueling calculations needed to work them out—under his supervision. Unfortunately the student hadn't finished the job. "He was not a terribly strong student," Kip said.

Stephen had signed up to speak at the Dublin conference when the calculation was far enough along that he was confident it would pan out. But the work to prove that it really would pan out was never completed—and Stephen, convinced of the answer, was not going to spend his limited time on the planet remedying that. And so his vague talk in Dublin, and his description of it in the published conference proceedings, are all that he ever said on the matter.

Stephen's talk and his concession made worldwide front-page news, but it was just a media show, much ado about not much. By the time he gave his talk virtually all physicists had already come to believe that the information wasn't lost—though no one, including Stephen, could prove it. And of those who hadn't come to that conclusion, no one, on the basis of Stephen's talk, joined him in changing their mind.

It amazed me to observe the power of labels. Before Stephen was known, strangers sometimes labeled him, based on his appearance, as defective both physically and mentally, and they looked away. Once he'd been declared a modern Einstein, however, the media would feast on whatever he said. Had the Dublin talk been given by a less celebrated figure, it would have been a fine addition to the swirl of ideas physicists toss around on the topic but it would have commanded no space in any newspaper. Since it was Stephen, though, the talk became a media circus.

For Stephen himself, his conversion was a momentous—even a joyous—occasion. Proving yourself wrong is not, for most people, cause to break out the champagne, but, like Zel'dovich when he'd finally understood Hawking radiation, Stephen's ultimate interest was always in the truth, and he rejoiced in the feeling of understanding something he hadn't realized before, something very important to physics.

Today, more than a decade and a half after the Dublin conference and well over forty years after Stephen's discovery of Hawking radiation, the minority supporting information loss is even smaller. Virtually all physicists believe, as Stephen said, that "if you jump into a black hole, your mass energy will be returned to our universe" in a form that, though mangled, "contains the information about what you were like."

But though we believe that the information is *not* lost, there is still no definitive explanation of what is really going on. There are so many theories—in addition to the scenario Stephen had presented—that physicists writing review articles don't list the individual theories, but rather the different *categories* of theory, each category containing many variants. Stephen himself stuck with his conversion to the majority point of view, but not necessarily to his original argument—he continued to work on new, alternative reasons for drawing that conclusion. He worked on it, on and off, until his dying day, and it became the topic of his last physics paper, published posthumously in 2018.*

* Haco, S., Hawking, S. W., Perry, M. J., et al. Black hole entropy and soft hair. *Journal of High Energy Physics,* (2018) 2018 :98.

II

It was Spring 2010. It had been five years since we started planning *Grand Design* and four years since we began to write it. I walked up the stairs at DAMTP and turned toward Stephen's office as I'd done so many times in those past years and every day for the past week. But this time was different. This day was ordained to be the day we'd finish the book.

After we'd missed deadline after deadline, the bigwigs at Bantam seemed to have lost patience. Without consulting us they'd scheduled the book for release and put it in their sales catalog. It was like when my son Nicolai was gestating. He seemed to be having a good time in the womb and stuck around well past his due date. The doctor finally put his foot down, scheduled a cesarean, and yanked him out. That's what Bantam was doing to us. We'd promised to complete the book in a year and a half, and now it had been four. It was time for the yank.

I could understand Bantam's point of view. If our manuscript were a child it'd be starting kindergarten soon. Human maturation is part of the miracle of life, but there's no such magic when it comes to writing a book. Had our manuscript come to term? Was it fully developed? I thought so, and people criticize me for being a perfectionist. But Stephen was still tin-kering. He had outperfectionisted me. Meanwhile book launch dates and events had been planned, marketing and PR activities scheduled. Sales reps had been touting the book, and bookstores had placed their orders. Things were happening that would be hard to reverse. Bantam was telling us, not so indirectly, that this time we had to deliver—or else.

Or else what, I didn't know. We were close to being done, but I hadn't seen any indication from Stephen that our deadline meant anything to him. He seemed immune to the "or else's" of the world. He'd had plenty of them, from his doctors, his body, his wives. One time, shortly after his split with Elaine, the media warned that there had been a bomb scare at a well-known night-club in London, the Tiger Tiger. People who wanted to stay in one piece stayed away. Stephen headed over and flirted with a woman at the bar. He was not a guy you were going to cow with dire warnings. So when he'd said he didn't care if it took us ten years to get the book right, I took him at his word. Whenever I recalled that, I got a knot in my stomach.

Stephen's office door was closed, so I wandered into Judith's office.

"Leonard!" she shouted. "A big day for you!"

She could tell from my reaction that I wasn't looking for-ward to it.

"Now, don't be discouraged. Look how close you are! You'll make it. You're almost there."

"Could you do me a favor and shoo people away today? Don't let them wander in."

"I shall guard the fortress!" she said. "But of course, if Ste-phen summons anyone, I can't stop that."

"You have your ways," I said.

Judith liked that I'd said that. She was proud that she had her ways.

After a few minutes Stephen's carer Cathy opened his door. "Leonard," she said, "he's ready for you."

When I'd first suggested to Stephen that we write a second book together based on his current work, it was the theory he'd invented in 2003, called top-down cosmology, that I had in mind.* After several chapters to lay the groundwork, we explained the crux of that theory in chapter six of *The Grand Design,* a chapter called "Choosing Our Universe." It was the most difficult chapter in the book. It wasn't the final chapter—as planned, there were two more that followed—but it was the chapter we were still working on, on the morning of that supposedly final day.

Top-down cosmology was an extension of Stephen's research in the 1980s on the no-boundary proposal. The goal of both was to examine the evolution of the universe from the perspective of quantum theory. They both, like his work on the information loss problem, relied on the Feynman sum over histories to calculate what quantum theory predicts.

As I've said, the sum over histories is usually applied to elementary particle physics, where the term "history" refers to the path a particle follows through space.† In top-down cosmology, as in the no-boundary proposal, the entire universe plays the role of the particle. As a result, while in more traditional calculations you had to take into account all possible particle trajectories, in his calculations Stephen had to consider all possible histories of the universe. In other words, to calculate the probability that the universe now has this property or that, he had to add up contributions from all possible ways the universe

---

* See Stephen Hawking, "Cosmology from the Top Down," *The Davis Meeting on Cosmic Inflation,* March 22–25, 2003.
† Technically, in quantum field theory, the sum is over field configurations.

might have evolved. That was an unusual approach—far different from anything Feynman had envisioned.

If one could do the math, the Feynman method would in principle shed light on any observation about the universe that you might make. But as usual, you cannot do the math. To make things more manageable Stephen considered a vastly simplified cosmic model that took into account only the gross structure of the universe. That made sense, because he wasn't interested in making predictions about individual atoms and molecules on earth or anywhere in the universe, but only about the universe's large-scale properties.

The top-down aspect of the work enters the picture because in ordinary cosmology, physicists assume some beginning for the universe and calculate how the universe evolves from that point forward in time. Stephen called that the "bottom-up approach." He didn't like that approach, because in his earlier work on the no-boundary proposal he'd concluded that the universe had had no single definite origin.

That conclusion was a reflection of one of the most famous and strangest aspects of quantum theory, that objects—in general—do not have definite properties, only probabilistic ones. For example, at some designated "zero hour," a quantum version of you might have a 50 percent chance of being in your downstairs kitchen and an equal probability of being in your upstairs bathroom. We physicists would say that your "initial condition" is neither in the kitchen nor the bathroom, but rather a "superposition" of those two states. According to the mathematics of quantum theory, the chances of you being elsewhere in the house at any later time would be influenced by both states in that initial superposition. Similarly, according to Stephen, the initial state of the quantum universe was a superposition of different possibilities, and we have to take all of them into account in order to understand our present. In Stephen's opinion, that made the bottom-up approach untenable.

Stephen argued that you should adapt Feynman's technique to include all possible origins of the universe. That means that the histories that enter into the Feynman sum depend only on the state of the universe at the present time. As Stephen liked to say, it means that the present determines the past, rather than the past determining the present. That's why he called his analysis "top-down" rather than "bottom-up."

Even with the simplifications he'd made, Stephen couldn't solve the equations that arose from his analysis. But he was able to determine some properties that the solution would have and extract some meaning from his model. He was fascinated by much of what he found. If his view is correct, it means that an infinite number of universes—a multiverse—appeared spontaneously, from nothing, with different futures ahead of them. This collection of universes is analogous to the different particle paths in the usual application of Feynman theory. The laws of nature observed in any given universe would depend on the history of that universe. There would be universes with every possible set of laws, universes in which a proton has the weight of a brick or in which gravity is so strong that a typical star burns itself out in just one of our years.

The mathematics of Stephen's theory suggests that in their early stages, most of those universes would expand but that the expansion would last only a short time. After that, the universes would collapse back into the hot dense ball from which they came. In those universes, there would be no time for galaxies and stars to form. But some universes, with just the right laws, would grow large enough to be safe from collapse.

Some of those universes could harbor life, and among those universes that could, some would. The creatures that emerge in those universes, if they are intelligent enough to decipher the laws of nature, would find that those laws have a very special form, one that allows them to exist. We are creatures of that special type, in a universe of that special type. That is the essence of what we were saying in *The Grand Design*.

After a career that spawned discoveries relating to the big bang singularity, the laws of black hole physics, and black hole evaporation (which, among other things, raised the information loss issue), the no-boundary proposal and top-down cosmology were Stephen's last grand initiatives. They were also his least influential. As is usual in work on the frontiers of physics, some colleagues were skeptical of the assumptions he'd made. Some were suspicious of his mathematical approximations. Some didn't understand his theory. And some simply found alternative theories more convincing.

To this day the jury is still out regarding both initiatives. Stephen proposed that analysis of the cosmic microwave background radiation might provide supporting evidence, but such an analysis will depend upon technology that doesn't yet exist. So, like most theories in modern cosmology, the no-boundary proposal and top-down cosmology are mathematically intriguing but difficult to test.

Most days, when Stephen got to the office in the late morning, he'd reply to a few emails and would read any articles of interest that had been posted on the ArXiv.org website. That's a repository for papers in physics, astronomy, and mathematics. Scientists in those fields usually post their results there at the same time they submit them to an academic publication. It can be months before an article appears in a journal, and reading the papers on the arXiv gives you a preview. On the other hand, the arXiv does no editing and only a cursory filtering. What you find there may need revision. It might not even warrant publication. It's reader beware, but Stephen checked it almost every day.

On this supposedly final day, however, Stephen had sacrificed his prelunch quiet time, and we'd started earlier than usual. As noon approached I was aware that we would soon be taking an hour-and-a-half break, so I was racing to finish our discussion of chapter 6. I wanted to move on to other issues after lunch. If we didn't, I feared that "Choosing Our Universe" would swallow the universe of our afternoon.

This was a Friday, and our official deadline was 8:00 p.m. Cambridge time. That equated to 3:00 p.m. New York time. I was grateful that the deadline wasn't earlier, but 3:00 seemed an odd choice. I imagined an old-fashioned newsroom like you'd see in the movies, with hordes of Bantam employees waiting for us, "holding the presses" until the last possible moment, at which time they would print a million copies and rush them to the newsstands. That's not what was going on, but it felt that way.

Among the issues our editor, Beth Rashbaum, had asked us to address in this final meeting was a simple addition to the way we introduced Feynman. She was our Peter Guzzardi, a big help in providing us with a nonscientist's perspective on our writing. Since Stephen and I had both known Feynman and he was an important presence in the book, she felt we should include a few words about him when we first mentioned him. I suggested something to Stephen, something about the broad and deep impact of Feynman's work in many areas of physics— Feynman's love of the entire field, not just the sexy frontiers that attract most people of his caliber.

Stephen responded with his *no* face and started typing his own description of the man. "Let's say: He was a colorful character who liked to play bongo drums at a strip joint near Caltech," he said. He smiled as his computer voiced his words.

I had to admit that he had captured Feynman's character as well as one can in a few words. It was unconventional, but it was short and sweet and true, and it was vintage Stephen. It could (almost) be a haiku, I thought:

He was a colorful character
who liked to play bongo drums
at a strip joint near Caltech

The book included plenty of information about Feynman's approach to physics, much of which was going to prove challenging to general readers, so why not set him up with a characterization that would disarm people before they got to the hard stuff? "Okay," I said. "Let's use that." And so we did, with a bit of editing.

Just then a tall, sturdy woman in her late thirties walked in. Her name was Diana, a former carer. She was the only person who got past Judith that day. She had in her hand a book she'd been reading to Stephen—Dickens's *A Tale of Two Cities*. Fitting, I thought, since for Stephen it seemed that both the best of times and the worst of times were never far off. Stephen had always loved books. He'd grown up with books, books that were strewn all over the house, books that were stuffed two deep into bookshelves everywhere. Even when he had friends over for dinner, his parents would sit at the table and read as they ate. Now Stephen was about to emulate them. "I thought I'd read to Stephen over lunch, if you don't mind," Diana said. It was nice of her to ask me, but it didn't matter what I said. Stephen's expression indicated that he welcomed her intrusion. I didn't mind, since it was difficult to communicate with him while he was eating anyway.

Diana often read to Stephen, sometimes for hours at a time. He especially loved the classics. Years later, when his health was starting to fail and it was time to choose the next book, someone recommended to Stephen that he stick to something short. Stephen chose *War and Peace*.

With Diana there, any hope of further delaying lunch was dashed. Cathy wheeled Stephen away from his desk and then the four of us left his office, heading down the hall and across the bridge to the next building, which housed the cafeteria. Before we went in, Cathy gave Diana a look. At first I was confused, but then I got the picture. Diana took charge of Stephen and the wheelchair, and walked into the cafeteria. Cathy stayed outside and pulled a pack of cigarettes from her purse.

I stuck with Cathy and bummed one off her. We each lit up. We stood beside a sign that said not to smoke there but we were outside, and so, I thought, bugger the sign.

"I hate cigarettes," I said.

"Who doesn't?" Cathy agreed, and took a big drag.

"No, really," I said. "I never smoked before I started this book with Stephen."

"I never smoked before something, too," she said. I looked at her, not sure what she'd meant by that. She took another drag and shrugged. "You're not alone," she told me. "Stephen can have that effect on people."

We smoked in silence for a few minutes. Then she said, "We'd better go rescue Diana."

"Does Diana need to be rescued?" I said.

"Actually, she'll have been happy to have had some time alone with Stephen," she said. "One doesn't often get Stephen alone. But he's taken to Diana. Some of the carers are jealous, but not me. He was lonely. People think all he needs is physics but that's malarkey. He's lucky to have her."

I felt that, too. It had seemed to me that Stephen wasn't quite the same after his split with Elaine. When he and Jane had broken up, it was because they'd grown apart. She'd found someone else, and he'd found Elaine. Although they had tried to hold it together in an unwieldy foursome, the marriage proclaimed one pairing and their hearts dictated another. So they'd split, like a molecule that breaks in two because one of its bonds is weak. Stephen's breakup with Elaine had been different. He'd split off from Elaine without a partner. He became a single atom, floating unbonded through the lonely vacuum of space.

The saying goes *Water, water, everywhere, but not a drop to drink,* and that's how it was with Stephen and companionship after Elaine was gone. There were carers, carers everywhere, as well as colleagues, and fans, and the media, but they supplied no intimacy for him. His best friends, Kip and Robert, lived outside England. He couldn't phone them and have a long talk, as the rest of us can with our friends. That left Stephen with no one to share his life with, no one to confide in, no one to be his, every day and every night. It seemed to me it would be hard enough to find new authentic relationships when you're famous, but in his condition it must have felt even bleaker. He still saw Elaine from time to time, and they had even gone on a few vacations together. Rumor had it that he'd actually talked to her about getting back together. But then he connected with Diana.

Like Elaine, Diana had started as one of his carers. Then she'd taken another job, but it didn't pay much, so he'd allowed her to move into the middle bedroom upstairs at his house. They shared a love of literature and music. She'd read him whatever he wanted, and she was a decent pianist. He bought her a piano, and she repaid him with long recitals.

As we walked into the cafeteria I spotted Diana with Stephen at the far end, opposite the food line. It was an enormous space with plenty of windows, much longer than it was wide. It had some height, too—a ribbed ceiling curved up from the sidewalls to perhaps twenty feet at its center. I couldn't decide whether it felt like the dining hall on the starship *Enterprise* or a Roman warship. I picked out a sandwich. Cathy was still unpacking the food she'd brought for Stephen, and Diana was reading to him.

Stephen's children didn't approve of Diana. She was thirty-nine years younger than Stephen and suffered from manic depression. But if Diana was troubled, it didn't seem to bother Stephen. Maybe he was used to it. Speaking of Elaine, he'd once said, "She's mixed up. But it's time I helped someone else. All my adult life people have been helping me." Was he attracted to troubled women? I wasn't sure. I thought Diana was intelligent and well read. Nice to chat with; I learned stuff. But that was when she was on her meds.

Some had suspicions that Diana was after Stephen's money. I found it sad if people assumed that money was all Stephen had to offer. I knew people had their reasons for thinking that, just as some might have had their reasons for questioning Elaine's love for Stephen. Stephen's physical condition couldn't have held much allure. But physical desire can result in love—or, conversely, it can be the result of love. Maybe that's what happened between Diana and Stephen. Some doubted it. Can one fall in love with someone who can't move or talk? How would that love develop?

It appeared to me that in Stephen, Diana had found a deep connection, not with his body but with his soul. "He had the most expressive face in the world," she told me once. "An eyebrow move here, a mouth twitch there—I know him so well I can follow his thinking. I could write a whole book just on how to communicate with Stephen."

I could feel the affection in Diana's voice when she said that. On another occasion she told me she wished she could trade places with him. She wished she could give him a gift of her health, leaving her the quadriplegic rather than him. Her eyes had teared when she said it, and I believed she'd meant it. Maybe I was a dope. Maybe Stephen was, too, like those gravity researchers Feynman had disdained at that Warsaw conference decades earlier. Maybe you had to be a dope to care about something as far removed as the origin of the universe, or to bond with an unsettled woman who devotes herself to you.

After lunch, back in Stephen's office, our work was going slowly. It had become tedious, with Stephen making numerous small changes here, there, everywhere. Some had to do with the artwork. For that, I had recruited a "futurist" digital artist, Peter Bollinger. Stephen and I both thought he'd done a great job. But now Stephen was asking to change some colors and rephrase some captions.

In my opinion those late alterations did not noticeably improve the book. They did no harm, either. They were just changes. For the most part I didn't argue, and I didn't mind making them. On the other hand, I was worried that we were approaching the finish line "asymptotically." That's a mathematical term meaning that you get closer and closer to something without ever reaching it, at least not in a finite amount of time. If each second you travel half the distance to a finish line, then you are approaching it asymptotically. That's fine in the mathematics world. In the business world it doesn't go over so well.

It's not that I thought the book was perfect. It was a reflection of our thoughts and ideas, and that certainly left room for debate. In particular, there was no avoiding the fact that we were writing about a theory that was still a work in progress. That meant there were some aspects of the theory that even Stephen didn't understand. Once I'd driven up to UC Santa Barbara to clarify a few points with Stephen's oft-time collaborator Jim Hartle, who had joined Stephen in working on top-down cosmology. He gave me a private mini-seminar on the relevant part of their work, and then I'd written about the issues in question. On my next trip to Cambridge Stephen told me that what I'd written was wrong. I was sure I'd understood Jim's lecture, so I was adamant. I started explaining my reasoning to Stephen, but he made a face and started typing a response. "Jim told you what we thought at the time," he said. "Since then we realized we were wrong."

Our lunch seemed to have energized Stephen, and we kept at the revisions all afternoon, punctuated only by the usual breaks for tea and vitamins, mashed banana, and for Stephen to "go on the couch." Five o'clock came. I mentioned the time. Stephen frowned. He didn't want to discuss it. Six o'clock came, then seven. We were working slowly, steadily, incrementally, as we had for the past four years, as if this day and the next hour were nothing special.

At 7:45 I gave up and stepped next door into Judith's office. She seemed to be working on some financial documents for Stephen. She did that sort of thing. She even helped him negotiate contracts and manage investments. I knew this because on occasion she'd asked my advice.

Judith pushed the papers aside and diplomatically laid a folder on top to shield them from me. It made sense but it didn't matter. Stephen had so little privacy and enough carers and other aides who were both curious and talkative that after four years I figured I knew more about his finances than about my own. I didn't care much about either, as long as I wasn't broke.

"Leonard!" Judith finally said with enthusiasm. A moment later my look told her that her enthusiasm had been misplaced. "Not going well?" she said.

"Good guess," I said.

I must have looked disturbed, like that art therapy patient she'd had back in Fiji when the topic of his father came up. The father whose head he'd chopped off.

"I'm sorry," Judith said. "I know how important the deadline was. I really did think you'd finish."

I'm sure I made another face, and I'm sure it wasn't pretty. I asked her to look into extending my room reservation at Caius and postponing my flight. I silently contemplated all the things I'd have to reschedule if I didn't go home when planned. I contemplated Bantam's reaction. I dreaded the coming confrontation. I'd been scheduled to leave the next afternoon, Saturday. I'd assumed we'd be finished, but I was now wondering how I could have been so stupidly optimistic. I left her to it and stepped back into Stephen's office.

As I entered Stephen brought up a new issue with the art. One of the illustrations was a triptych meant to depict an idea that comes up in string theory/M-theory. Each of the three triptych panels showed a futuristic-looking pink-and-blue drink with a straw standing in the glass. The panel on the left showed it up close. The middle one was a view from farther off. The rightmost panel was a distance shot. In that view, the straw looked like a line rather than the hollow cylinder one could make out in the other panels. The idea was that a higher-dimensional object like the straw can appear to be one-dimensional—i.e., a line, if you view it from a distance.

Stephen seemed to have a problem with the illustration. I tried the twenty-questions method to guess his issue but had no success. He cut me short and started typing. It was 7:59. I thought about how I liked the triptych. It made an important technical point in a simple manner. I couldn't imagine what his objection might be. Finally he finished, and his computer voiced his complaint. "The straw in the far right panel is too long," he said.

I felt like having a smoke. I felt like crying. I looked on Stephen's desk for the illustration he was referring to, but I didn't see it anywhere. He must have been speaking from memory. I got the proofs of the illustrations from my briefcase and found the one he was talking about. He was correct. If, in the far right panel, you bothered to carefully compare the length of the straw to the height of the glass, you'd realize that the straw was unrealistically long. I'd looked at it a dozen times and not noticed. Just then I hated Stephen. What the hell was wrong with him? Who has a mind like that?

I tried to calm myself. I took a deep breath. I made a note to tell the artist about the straw. I said to Stephen, "Okay. Next issue?" I tried to keep the tension out of my voice. He frowned. What was he frowning about? I had no idea. He typed something.

"It is 8:00," he said. "We are finished."

I hadn't seen that coming. I babbled something, I don't know what. But I do remember Stephen's response.

"I need a strict deadline," he said. "Otherwise I never finish."

He put on a big smile. I stared at him. I thought I ought to reply, but I was speechless. He kept typing.

"The book is good," he said. "Thanks for writing it with me. Let's have dinner."

At Stephen's house dinner was still simmering, stewing, or whatever. As his carer worked on completing it, Diana sat at the piano, playing a classical piece, and Stephen sat nearby, taking it in. I didn't recognize the music, but I did recognize the wine closet. I walked over and pulled out an unopened bottle of cognac. There wasn't much hard liquor in there; I never saw Stephen drink any. But there was this cognac. I assumed that, like the wine, it had been a gift.

I didn't feel like interrupting Stephen to ask if I could open the bottle, so I just helped myself. I took one of the wineglasses from the dining room table and gave myself a little splash, like you do with cognac. Then I thought better of it and let it flow. It was a wineglass, so why not fill it as if I were pouring wine? I didn't ask if Stephen wanted any. I knew he didn't, just as I knew that I did. Then I moved a chair over to sit with him near the piano.

Diana played with soul. She could play softly, but at the appropriate places she also played with passion, even anger. She could really pound away at the angry parts. At one point she hit a sour note but played through it. "I prefer to hear you play with mistakes than to hear it perfect on the radio," Stephen had once told her. He may have meant he preferred the live sound to what came from his fancy speakers, but I didn't take it that way. I took it as a declaration of love. Stephen had loved Jane, and he'd loved Elaine. As insightful as he was in physics, he probably had no idea what had happened to all that love. Was it still there, somewhere? Or had it evaporated? Pondering those questions could be enough to make you cynical, but Stephen didn't seem to be. He seemed to be in love for a third time.

Watching Diana look at Stephen, it was clear to me that the love was mutual and that she loved him for the right reasons. Stephen appeared to feel that, too. Diana was at times a troubled woman, but in the years she had lived in Stephen's house, she had been better. Stephen knew of her mental health issues, and he accepted her just as she accepted him. She had started as Stephen's carer, but she felt that he was also *her* carer. She liked the way he'd ask her to hold his hand, to kiss his cheek, or to lie in bed with him. She liked going with him to get their hair done together. She liked taking him to Wimpole Home Farm to watch the pigs feed. He was fascinated by cosmology, but he was also fascinated by pigs. Diana thought his ancestors on one side may have been pig farmers. She should know: at this point in Stephen's life, she knew him as well as anyone.

I was near the end of my drink. The cognac had felt warm and soothing going down. I liked the rush and relaxation it provided. I liked that I felt comfortable here in Stephen's home. I liked the love I sensed in the room. I liked that I didn't have to go anywhere for a while, especially because walking any distance at this point would have been a challenge. Most of all, I liked that we were done with the book.

But that happiest of feelings also had a flip side. As I watched Stephen luxuriate in the music, I knew this was the end of an era in my life, that I'd not see him often in the coming years. Stephen and I had grown close, but when would our paths cross again? After being in the trenches together for years, writing two books, arguing, cooperating, sharing our meals and our thoughts, would our connection now fade, like that with his ex-spouses? There would be his yearly visits to Caltech, and once in a blue moon, maybe, I'd find myself in England, but after the intense connection of the past few years that scenario seemed only a pale reflection of what I'd had with Stephen.

Those thoughts were now making me sad. How could my elation over finishing be so short-lived and easily overshadowed? I felt like I needed another drink, but the dining room table seemed too far to travel for it. I sat back and joined Stephen in focusing on Diana's music. I wish now that I'd asked her what it was.

Epilogue

Bantam published *The Grand Design* in September 2010. The artist never fixed the too-long straw, but it didn't matter. On the morning of September 2, I was walking my daughter Olivia to school when my cellphone rang. It was Judith, and she was agitated. "Leonard!" she shouted. "We need your help!" I had no idea what she was talking about.

"Haven't you seen *The Times?*" she asked.

"The New York Times?" I said. Yes, I'd read it.

"Not *The New York Times,*" she shouted. *"The* London *Times!* Haven't you seen it?"

"Judith, who here reads the London *Times?*" I asked.

"Well, Google it and look at the headline! It says 'Hawking: God Did Not Create Universe.' It's created a furor!"

"That's wrong," I said. "We said God isn't *necessary* for creating the universe, not that physics proves he didn't."

"Well, the press is all over it, and Stephen can't handle it all. We need your help! You need to take the interviews."

And so it began. We knew the book would get some attention, but we had no idea how much. You know you've reached people when a physics book gets talked about on ESPN and in *Men's Health* magazine.

Grand Design had obviously captured people's imagination, but though most of the reaction was positive, we were also vehemently denounced in some quarters. Apparently provoked by our views on creation, some of those attacks were personal. People who had little idea of what Stephen was really like assumed that they understood his motives. Stephen was accused of scheming to use his disability as a marketing tool, and of trying to profit by attacking God. Stephen just smiled at the criticisms. I suppose that on an annoyance scale ranging from one to "I can't move anything below my neck," those ad hominem attacks scored pretty low.

In 2013 Stephen asked Diana to marry him. She had long since moved out of his house, but they had remained as close as ever. The proposal came one night after dinner. Stephen began by saying, "I cannot get down on one knee," and then went on to declare his love for her and to ask if she would be his wife. Soon after, on her birthday, they went to a jeweler and chose a ring together. Then they went to a restaurant to celebrate over dinner.

But the marriage never happened. Stephen craved Diana's companionship, but his desire for family harmony proved even more powerful, and it seemed his children couldn't get past their issues with Diana, or their more general concern about gold diggers. I didn't agree with that assessment, but I knew it came from their love for him.

When Stephen married Elaine, of his children, only his son Robert—who lived in Seattle—attended the wedding. During their years of marriage, there continued to be tension between Elaine and his other two children, both of whom lived nearby. It had at times made family occasions awkward and been unpleasant for both Elaine and Stephen. Stephen didn't want more of the same, so he backed out of his proposal to Diana. Though she was devastated, they stayed dear friends. He asked her to keep the ring. She still has it.

In time the commotion over the book gradually faded. So, alas, did my contact with Stephen. As I'd feared, as the years passed it proved difficult to keep up our connection. His computer assistant, Sam, moved on. Joan, his oldest, most faithful carer, died. Judith retired. I didn't know Stephen's new personal assistant, so a critical source of news and communication had vanished. Stephen and I would email on occasion, but the couple of times I'd been in England, he'd been away. And so I saw him only during his yearly sojourns at Caltech. After 2013 even those ceased due to his declining health.

In my final visit with Stephen I dropped by the house in Pasadena where he was staying and spent a Sunday with him, Kip, and a few others who came and went. One was Buzz Aldrin, the former astronaut, the second person to walk on the moon. It was a lazy afternoon, with a little food and a lot of idle chatter. Stephen's communication speed had dwindled by then to less than a word a minute. As a result, even simple sentences now took five to ten minutes to emerge. But Stephen could still smile and growl. And so, on that last afternoon together most of our conversation was a big game of twenty questions. But not all our conversation. At one point we—mostly I—reminisced a bit about our work on *Grand Design*. I drifted to the subject of his career and asked him which, of his many discoveries, accomplishments, creations, had been his favorite. A few minutes passed before the answer came. When it did, what he said was "My children."

I was staring at a computer screen when the news flash came: *Stephen Hawking is dead.* He died at his home on Wordsworth Grove on March 14, 2018. It had been more than four years since I had seen him. Kip had last seen him the previous November, Robert Donovan that past December. He said that though Stephen hadn't appeared particularly sick, he seemed to be expecting death. He'd hired a lawyer to get his affairs in order.

I'd expected to see Diana at Stephen's funeral, but she wasn't there. As it turned out, she hadn't been put on the family's guest list. The interment occurred a few months later. She wasn't at that, either. That morning she traveled to where it was to be held—Westminster Abbey. She got there early and attended the morning mass at 7:30. That was a public event that, as usual, drew a couple of dozen of the faithful. The interment, which would be attended by over a thousand people, came a few hours after it. By then there were guards, armed police, and various officials there to usher the invited in and to keep the uninvited out.

Again Diana wasn't on the guest list, so she was barred from entering. She pled her case to one of the officials, but he denied her. "We have to keep this dignified," he said. And so she stood, shut out, among the crowd gathered outside. She felt lost in a sea of strangers, none of whom had read to Stephen for hours, wiped the sweat from his brow, worn his engagement ring, or lain in bed with him, holding him in her arms. As she strained to hear a trace of sound emanating from the service inside, she felt her sadness mingle with an immense feeling of rejection.

When the interment was over, the crowd dispersed but Diana lingered. Stephen's friend Neil Turok spotted her as he was walking out of the church. He approached her, vouched for her, and walked back inside with her. She cried as she paid her last respects.

Like Stephen's other loves, Diana was a woman of deep faith. He once told her, "Religion is for those afraid of the dark." He didn't mean to insult her; he was just being roguish. She replied that everyone is afraid of the dark. He accepted that, at least as an approximation. And when he died, she took comfort in her faith. "I have to believe I'll see him again," she said. "I can't believe there is nothing outside the universe. The universe couldn't be so cruel. I look forward to the time when I'll join him, wherever that is."

People sometimes ask me how Stephen could have won the battle, for so many decades, over the hopelessness that is natural when your death continually seems imminent. I answer that faith was his greatest weapon. He may not have had faith in a god, but he had faith in himself. He had faith, when he went to bed each night, that he would wake up in the morning. He had faith that when he entered the hospital, he would come out; that despite his doctor's orders he could travel the world and survive. He had faith that those who loved him loved him for who he was and not for his money or fame. He had faith that his continued life would bring more reward each day than the punishment he endured each night, with his restless and painful sleep, and the indignity of having to be spoon-fed and bathed by others.

After Stephen died I dug up some of the material I had saved from our time together—old notes, rough drafts of chapters I had printed out, annotated with the comments Stephen gave me. I ran across the get-well card he had sent after that brush with death I'd had in the hospital. I remembered his concern and the strength I had felt simply by thinking of how he had so regularly weathered equally threatening storms.

I missed Stephen. We'd been through much together, and it had left me a better person. We never spoke about his philosophy of life. But knowing him and sharing parts of his life reinforced my belief in my own hopes and dreams, and in my ability to fulfill them despite the hardships that will inevitably come my way, as they come to all of us. Often we limit our chances at success by limiting the goals toward which we strive. Stephen never did that. Even the example he set by simply showing up at the office each day influenced me. It made me more tolerant of the problems in my own life, and more grateful for all that is good in my life, however small.

We can get used to anything, and we can accomplish, if not anything, then at least much more than we give ourselves credit for. To grow close to Stephen was to understand this, and to realize that we need not wait for a debilitating disease to inspire us to make the most of our time on earth. And so I continue to do physics and to write my books.

To those who knew Stephen from afar it could appear that, for him, just to live was to climb Mount Everest. After I got to know him it struck me that he *was* Mount Everest. An immovable giant, immune to the passage of time and able to withstand even the most violent storms nature hurls at it.

I know that time eventually slays us all, but to observe the power Stephen exerted over his life gave me the feeling that he could also control the timing of his demise. Hearing of Stephen's passing, I couldn't help but believe that death had not overpowered Stephen, but rather that he'd simply decided to cease repelling its attacks. He'd done enough, seen enough, lived a good life full of friends and children and love, and physics. He'd found meaning in his life, and even in his suffering, which inspired him to help others with similar needs. And so, at the end, he said his goodbyes, and when his disease again came at him, he lay down his arms and rested in peace.

A Note on Sources

I based this book upon my own experiences, supplemented with interviews with fifteen of Stephen's close friends, carers, and colleagues, each lasting from ninety minutes to eight hours: Sam Blackburn, Bernard Carr, Judith Croasdell, Robert Donovan, Diana Finn, Peter Guzzardi, James Hartle, Elaine Hawking, Don Page, Martin Rees, Vivian Richer, Erhard Seiler, Kip Thorne, Neil Turok, and Radka Visnakova.

I also drew background material from two biographies: Kitty Ferguson, *Stephen Hawking: His Life and Work* (London: Transworld, 2011), and Michael White and John Gribbin, *Stephen Hawking: A Life in Science* (New York: Pegasus, 2016). For some of the details of Stephen's life in the 1970s and '80s, I looked to Jane Hawking, *Music to Move the Stars* (London: Pan, 2000), and Kip Thorne, *Black Holes and Time Warps* (New York: Norton, 1994). Finally, I also gleaned a few details from David H. Abramson, "Saving Stephen Hawking," *Harvard Magazine* (May 9, 2018); Judy Bachrach, "A Beautiful Mind, an Ugly Possibility," *Vanity Fair* (June 2004); and Bernard Carr, "Stephen Hawking: Recollections of a Singular Friend," *Paradigm Explorer* (2018/1), 9–13. Generally, I used these references only as background or for looking up facts or as a source for quoting some things Stephen said.

Acknowledgments

In the years just before and after Stephen's death I dismissed out of hand the idea that I write his biography. There has been plenty written about him and I didn't want to rehash things that had already been covered. Then one day my agent, Susan Ginsburg, told me that my editor, Edward Kastenmeier, wanted to discuss the issue. I had already declined the requests of other publishers, and planned to decline this one, but when Edward called he surprised me by asking instead if I would write a memoir. Though the idea of writing something personal had appealed to me, I'd doubted that anyone would be interested. Yet if he was misguided enough to want to publish this, I was enthusiastic about writing it. And so we moved forward. I am grateful to Edward and Susan, not only for believing in this project but also in helping me form its vision, and for their brilliant guidance along the way. And to Donna Scott, my wife and in-house editor, whose advice was as incisive as theirs.

I am also grateful to my friends and colleagues who read and commented on the technical portions of this book. Theoretical physics is a beautiful but difficult subject practiced by passionate people—for without the passion it is hard to muster the patience and perseverance necessary to make progress. And so, for stealing time from their pursuit of new discoveries, thanks to Todd Brun, Daniel Kennefick, Don Page, Sanford Perliss, Erhard Seiler, Kip Thorne, and Neil Turok. I am equally grateful to those who read and provided advice with regard to the human side of the manuscript: Rob Berg, Catherine Bradshaw, Judith Croasdel, Casiana Ionita, Nathan L. King, Ceclia Milan, Alexei Mlodinow, Nicolai Mlodinow, Stanley Oropesa, Beth Rashbaum, Fred Rose, Julie Sayres, Peggy Boulos Smith, Martin J. Smith, Andrew Weber, and Mariana Zahar. Most of all, I owe a debt to Stephen Hawking for choosing to work with me, and for the warmth and friendship we shared over the years we knew each other. His passing has left a black hole in the lives of all who were his friends.

Leonard Mlodinow is a theoretical physicist and was on the faculty of the Max Planck Institute and the California Institute of Technology. His books have been translated into more than thirty languages and have sold more than a million copies. They include the best sellers *Subliminal* (winner of the PEN/E. O. Wilson Literary Science Writing Award), *The Drunkard's Walk* (a *New York Times* Notable Book), *War of the Worldviews* (with Deepak Chopra), *The Grand Design* (with Stephen Hawking), and *A Briefer History of Time* (with Stephen Hawking), as well as *Elastic, The Upright Thinkers, Feynman's Rainbow,* and *Euclid's Window.* He has also written for the television series *MacGyver* and *Star Trek: The Next Generation.*

A NOTE ON THE TYPE

This book was set in a version of the well-known Monotype face Bembo. This letter was cut for the celebrated Venetian printer Aldus Manutius by Francesco Griffo, and first used in Pietro Cardinal Bembo's *De Aetna* of 1495.

The companion italic is an adaptation of the chancery script type designed by the calligrapher and printer Lodovico degli Arrighi.

Typeset by Scribe,
Philadelphia, Pennsylvania

Printed and bound by Berryville Graphics,
Berryville, Virginia

Designed by Soonyoung Kwon